DETERMINANTS AND MATRICES

UNIVERSITY MATHEMATICAL TEXTS

GENERAL EDITORS
ALEXANDER C. AITKEN, D.Sc., F.R.S.
DANIEL E. RUTHERFORD, Dr. Math.

Other volumes in preparation

DETERMINANTS
AND MATRICES

BY

A. C. AITKEN, M.A., D.Sc., F.R.S.
PROFESSOR OF MATHEMATICS IN THE UNIVERSITY OF EDINBURGH

FIFTH EDITION

OLIVER AND BOYD
EDINBURGH AND LONDON
NEW YORK: INTERSCIENCE PUBLISHERS, INC.
1948

First Edition . . . 1939
Second Edition . . . 1942
Third Edition . . . 1944
Fourth Edition, Revised . 1946
Fifth Edition . . . 1948

PRINTED AND PUBLISHED IN GREAT BRITAIN BY
OLIVER AND BOYD LTD., EDINBURGH

CONTENTS

▼

CONTENTS

CHAPTER VI

SPECIAL DETERMINANTS: ALTERNANT, PERSYMMETRIC, BIGRADIENT, CENTROSYMMETRIC, JACOBIAN, HESSIAN, WRONSKIAN

CHAPTER I

DEFINITIONS AND FUNDAMENTAL OPERATIONS OF MATRICES

1. Introductory

THE notation of ordinary algebra is a convenient system of shorthand, a compact and well-adapted code for expressing the logical relations of numbers. The notation of matrices is merely a later development of this shorthand, by which certain operations and results of the earlier system can be expressed at still shorter hand. The rules of operation are so few and so simple, so like those of ordinary algebra, the notation of matrices is so concise yet so flexible, that it has seemed profitable to begin this book with a brief account of matrices and matrix algebra, and to derive the theory of determinants by the aid of matrix notation, in an order suggested by a naturally alternating development of both subjects. This first chapter is devoted to explaining the code. The reader is invited to take it with due deliberation, to invent at all times and verify examples for himself, especially in regard to the transposition of matrix products and to the multiplication of partitioned matrices, and, at first, to write out results both in ordinary and in matrix notation for comparison. The confidence and facility acquired by such practice will prove to be of constant service during the study of the later chapters.

2. Linear Equations and Transformations

The theory of matrices and determinants originates in the necessity of solving simultaneous linear equations and of dealing in a compact notation with *linear transformations*

A

from one set of variables to a second set. In the very early stages of elementary algebra we meet simple equations of the first degree, of the form

$$ax = h. \qquad . \qquad . \qquad . \qquad . \quad (1)$$

At later stages we meet simultaneous equations in two unknowns,

$$\begin{aligned} a_1x + b_1y &= h_1, \\ a_2x + b_2y &= h_2, \end{aligned} \qquad . \qquad . \qquad . \quad (2)$$

in three unknowns,

$$\begin{aligned} a_1x + b_1y + c_1z &= h_1, \\ a_2x + b_2y + c_2z &= h_2, \\ a_3x + b_3y + c_3z &= h_3, \end{aligned} \qquad . \qquad . \quad (3)$$

and so on. The method of solving by successive eliminations, and conditions under which unique solutions exist, may perhaps be known to the reader.

At a still later stage, in co-ordinate geometry of two dimensions, we encounter various linear transformations, such as

$$\begin{aligned} x &= x' \cos\theta - y' \sin\theta, \\ y &= x' \sin\theta + y' \cos\theta, \end{aligned} \qquad . \qquad . \quad (4)$$

representing a change of rectangular axes by rotation about the origin through an angle θ, and in the three-dimensional analogue of this we meet with

$$\begin{aligned} x &= l_1x' + l_2y' + l_3z', \\ y &= m_1x' + m_2y' + m_3z', \\ z &= n_1x' + n_2y' + n_3z', \end{aligned} \qquad . \qquad . \quad (5)$$

where the l_i, m_j, n_k are direction cosines. Indeed everywhere in mathematics we are confronted with equations of linear transformation; and this in itself is enough to justify the search for a code and a calculus.

The general set of m simultaneous equations in n unknowns is

$$\begin{aligned} a_{11}x_1 + a_{12}x_2 + \ldots + a_{1n}x_n &= h_1, \\ a_{21}x_1 + a_{22}x_2 + \ldots + a_{2n}x_n &= h_2, \\ \vdots \qquad\qquad\qquad & \\ a_{m1}x_1 + a_{m2}x_2 + \ldots + a_{mn}x_n &= h_m, \end{aligned} \qquad . \quad (6)$$

and the general linear transformation expressing m variables $y_1, y_2, \ldots, y_m$ as linear functions of n variables $x_1, x_2, \ldots, x_n$ is the same as this in form, but with the variables y replacing the constants h on the right of the equations (6).

The number m of variables y has been taken as possibly different from the number n of variables x. Such transformations of unequal numbers of variables can easily arise in practice, for example in questions involving perspective drawing, where a three-dimensional object may have to be represented on a two-dimensional sheet of paper. Assigned points in the object have three co-ordinates, the representative points on the paper have two coordinates only, and the representation is given algebraically by a set of equations in which $m = 2$, $n = 3$.

3. The Notation of Matrices

It would be intolerably tedious if, whenever we had occasion to manipulate sets of equations or to refer to properties of the coefficients, we had to write either the equations or the scheme of coefficients in full. The need for an abbreviated notation was early felt, and in the last century Cayley and other algebraists of the time made use of contracted notations such as

$$\begin{bmatrix} a_{11} & a_{12} & \cdots & \cdots & a_{1n} \\ a_{21} & a_{22} & \cdots & \cdots & a_{2n} \\ \cdot & \cdot & \cdots & \cdots & \cdot \\ a_{m1} & a_{m2} & \cdots & & a_{mn} \end{bmatrix} \begin{bmatrix} x_1 \\ x_2 \\ \vdots \\ x_n \end{bmatrix} = \begin{bmatrix} y_1 \\ y_2 \\ \vdots \\ y_m \end{bmatrix} \quad (1)$$

for a set of linear equations, detaching the rectangular scheme of coefficients a_{ij} from the variables x_j to which they referred. Later Cayley, by regarding such a scheme of ordered coefficients as an *operator* acting upon the variables $x_1, x_2, \ldots, x_n$ in much the same way as a acts upon x to produce ax, and by investigating the rules of such operations, formulated the algebra of matrices, meaning by matrices those schemes of detached co-efficients considered as operators. The theory was at first confined to square matrices, but the inclusion of general rectangular matrices increases greatly the scope and convenience of application.

Definition. A scheme of detached coefficients a_{ij}, set out in m rows and n columns as on the left of (1), will be called a *matrix* of order m by n, or $m \times n$. The numbers a_{ij} are called *elements* (by some writers *constituents*, by others *coordinates*) of the matrix, a_{ij} being the element in the ith row and j^{th} column. The row-suffix i ranges over the values 1, 2, ..., m, the column-suffix j over the values 1, 2, ..., n. The matrix as a whole will be denoted by A or by $[a_{ij}]$; or on occasion will be written out in full array. The element a_{ij} will often be called the $(i, j)^{th}$ element of A.

When once the rules are found by which matrices A and B can be added, subtracted, multiplied and divided, a proper sense being given to these operations through study of the laws obeyed by linear transformations, the materials requisite for an algebra will be available. This algebra, the algebra of matrices, has a very close resemblance to the algebra of ordinary numbers ; but it is a more general algebra, and the reader must be on guard, at first, against carrying over into it mechanically all the acquired manipulative routine of ordinary algebra.

4. Matrices, Row Vectors, Column Vectors, Scalars

A matrix may possibly consist of a single row, or of a single column, of elements. For example in **3** (1) we see on the left a column of elements x_1, x_2, ..., x_n, in fact a matrix of order $n \times 1$, and on the right a column of elements y_1, y_2, ..., y_m, a matrix of order $m \times 1$. Matrices of single row or single column type are of very common occurrence, and the general matrix itself may be viewed (**11**, Ex. 2) as an array of juxtaposed rows, or of juxtaposed columns. It is therefore convenient to distinguish row and column matrices by a special name and notation. We shall call them *vectors*, or more precisely *row vectors* and *column vectors*, and we shall denote them by small italic letters, the order, such as $1 \times p$ or $n \times 1$, being always understood from the context. For example, the two column vectors in **3** (1) will be written as x and y; and a device will be given later (**8**) for distinguishing a row vector from the column vector having the same elements. On occasion vectors may be written in full in row or column form with square brackets; but often, to economize in vertical space on the page, it will be convenient to write the elements of a column vector in horizontal alignment and to indicate, by the use of curled instead of square brackets, that a vertical alignment is intended. For example, we shall write column vectors as

$$x = \{x_1 \ x_2 \ \dots \ x_n\}, \quad y = \{y_1 \ y_2 \ \dots \ y_m\}, \qquad . \qquad (1)$$

and row vectors as

$$u = [u_1 \ u_2 \ \dots \ u_p], \quad v = [v_1 \ v_2 \ \dots \ v_q]. \qquad . \qquad (2)$$

In every situation in which vectors are being used it is essential to keep in mind, to visualize as it were, what kind of vector, whether row or column, is in question, and on no account to confuse the two kinds.

The matrix of order 1×1, that is, of one row and one column, is a single element. It will be found that the laws of operation of such matrices are in exact correspondence with those of ordinary numbers used as multipliers. We shall therefore identify such matrices with ordinary numbers or *scalars*.

To sum up, the matrix $A \equiv [a_{ij}]$ is the scheme of detached coefficients in some actual or possible linear transformation. It is not inert, but is to be imagined as an operator. It is also to be regarded as a complete entity, like a position in chess. If, for example, we interchange any of its rows, or its columns, what we obtain by so doing is in general a different matrix. Two matrices A and B are considered to be equal only when they are of the same order $m \times n$, and when all corresponding elements agree, that is to say when $a_{ij} = b_{ij}$ for all i, j.

We proceed to develop the algebra of matrices; but we shall find, when we come to consider what can be meant by the *division* of one matrix by another, that we are forced to turn aside and to study, define and evaluate a related set of numbers, namely the *determinants* corresponding to square matrices.

5. The Operations of Matrix Algebra

Addition. Consider for illustration the transformations

$$y_1 = a_{11}x_1 + a_{12}x_2 + a_{13}x_3,$$
$$y_2 = a_{21}x_1 + a_{22}x_2 + a_{23}x_3, \qquad (1)$$

and

$$z_1 = b_{11}x_1 + b_{12}x_2 + b_{13}x_3,$$
$$z_2 = b_{21}x_1 + b_{22}x_2 + b_{23}x_3, \qquad (2)$$

and suppose that new variables w_1 and w_2 are introduced by adding the corresponding y_i and z_i, thus:

$$w_1 = y_1 + z_1, \quad w_2 = y_2 + z_2. \qquad (3)$$

Then we have at once

$$w_1 = (a_{11}+b_{11})x_1 + (a_{12}+b_{12})x_2 + (a_{13}+b_{13})x_3,$$
$$w_2 = (a_{21}+b_{21})x_1 + (a_{22}+b_{22})x_2 + (a_{23}+b_{23})x_3. \qquad (4)$$

The process of obtaining (4) from (1) and (2) may logically be regarded as the addition of linear transformations, and may evidently be extended to the case of two sets of m equations, having the same n variables x_i on the right. The rule of *Matrix Addition* is thus suggested:

Addition of Matrices. *To add together two matrices A and B of the same order $m \times n$, we add their corresponding elements, and take the sums as the corresponding elements of the sum matrix, which is denoted by $A+B$. In symbols,*

$$A+B \equiv [a_{ij}] + [b_{ij}] = [a_{ij}+b_{ij}]. \qquad (5)$$

The rule can now be extended step by step (or could have been postulated at once) to give the sum, in the sense just described, of any finite number of matrices of the same order $m \times n$, thus :

$$A + B + \ldots + K = [a_{ij} + b_{ij} + \ldots + k_{ij}]. \qquad . \qquad (6)$$

For example,

$$\begin{bmatrix} 3 & -4 & 2 \\ -1 & 0 & 5 \end{bmatrix} + \begin{bmatrix} 2 & 6 & -1 \\ 4 & -3 & 2 \end{bmatrix} = \begin{bmatrix} 5 & 2 & 1 \\ 3 & -3 & 7 \end{bmatrix}.$$

Scalar Multiplication. Taking again equations (1), let us suppose that the scale of measure of y_1 and y_2 is altered in the ratio $1 : \lambda$, by the introduction of new variables $z_1 = \lambda y_1$, $z_2 = \lambda y_2$. It follows that

$$\begin{aligned} z_1 &= \lambda a_{11} x_1 + \lambda a_{12} x_2 + \lambda a_{13} x_3, \\ z_2 &= \lambda a_{21} x_1 + \lambda a_{22} x_2 + \lambda a_{23} x_3. \end{aligned} \qquad . \qquad (7)$$

This operation of multiplying variables by a constant scale-factor may properly be called *Scalar Multiplication*, and the rule for it is evidently this :

To multiply a matrix A by a scalar number λ, we multiply all elements a_{ij} by λ. In symbols,

$$\lambda A = \lambda [a_{ij}] = [\lambda a_{ij}]. \qquad . \qquad (8)$$

Linear Combination of Matrices. Combining now the rule of addition with that of scalar multiplication we have the rule for linear combination, with scalar coefficients, of any finite number of matrices A, B, ..., K of the same order $m \times n$:

$$\alpha A + \beta B + \ldots + \kappa K = [\alpha a_{ij} + \beta b_{ij} + \ldots + \kappa k_{ij}], \qquad . \qquad (9)$$

where α, β, ..., κ are scalar numbers.

Null or Zero Matrix. At this stage we can introduce the null or zero matrix, defined by the particular linear combination $A - A$ and denoted by 0. Whether rectangular or square, vector or scalar, it is seen to have *all* its elements zero. Thus for matrices A and 0 of the same order we have $A + 0 = A$.

6. Matrix Multiplication. Pre- and Postmultiplication

The simplest homogeneous linear transformation is of the form $y = ax$, an ordinary multiplication, and when in elementary algebra we have two of these such as $y = ax$, $z = by$ to perform in succession, the result may be written as the single transformation $z = bax$, the coefficients b and a of the separate transformations being combined in product to give the coefficient ba of the resultant transformation. When we perform successive linear transformations like this not on one but on many variables at a time we shall indulge in a natural extension of language and we shall say that the transformations are *multiplied* together in a certain order; we shall further call the matrix of the single resultant transformation the *product* matrix BA, by analogy with ba. To discover the appropriate rule for the elements of BA, let us consider the two transformations,

$$
\begin{aligned}
y_1 &= a_{11}x_1 + a_{12}x_2 + a_{13}x_3, \\
y_2 &= a_{21}x_1 + a_{22}x_2 + a_{23}x_3,
\end{aligned}
\qquad
\begin{aligned}
z_1 &= b_{11}y_1 + b_{12}y_2, \\
z_2 &= b_{21}y_1 + b_{22}y_2, \\
z_3 &= b_{31}y_1 + b_{32}y_2, \\
z_4 &= b_{41}y_1 + b_{42}y_2.
\end{aligned}
\qquad (1)
$$

These, as may be seen at once by substituting for the y_i in terms of the x_k, yield the single resultant transformation

$$
\begin{aligned}
z_1 &= (b_{11}a_{11} + b_{12}a_{21})x_1 + (b_{11}a_{12} + b_{12}a_{22})x_2 + (b_{11}a_{13} + b_{12}a_{23})x_3, \\
z_2 &= (b_{21}a_{11} + b_{22}a_{21})x_1 + (b_{21}a_{12} + b_{22}a_{22})x_2 + (b_{21}a_{13} + b_{22}a_{23})x_3, \\
z_3 &= (b_{31}a_{11} + b_{32}a_{21})x_1 + (b_{31}a_{12} + b_{32}a_{22})x_2 + (b_{31}a_{13} + b_{32}a_{23})x_3, \\
z_4 &= (b_{41}a_{11} + b_{42}a_{21})x_1 + (b_{41}a_{12} + b_{42}a_{22})x_2 + (b_{41}a_{13} + b_{42}a_{23})x_3.
\end{aligned}
\qquad (2)
$$

It is natural, therefore, to regard the matrix of the transformation (2), namely,

$$
\begin{bmatrix}
b_{11}a_{11} + b_{12}a_{21} & b_{11}a_{12} + b_{12}a_{22} & b_{11}a_{13} + b_{12}a_{23} \\
b_{21}a_{11} + b_{22}a_{21} & b_{21}a_{12} + b_{22}a_{22} & b_{21}a_{13} + b_{22}a_{23} \\
b_{31}a_{11} + b_{32}a_{21} & b_{31}a_{12} + b_{32}a_{22} & b_{31}a_{13} + b_{32}a_{23} \\
b_{41}a_{11} + b_{42}a_{21} & b_{41}a_{12} + b_{42}a_{22} & b_{41}a_{13} + b_{42}a_{23}
\end{bmatrix}
\qquad (3)
$$

as the *product*, in the sense just described, of the matrices

$$B = \begin{bmatrix} b_{11} & b_{12} \\ b_{21} & b_{22} \\ b_{31} & b_{32} \\ b_{41} & b_{42} \end{bmatrix} \text{ and } A = \begin{bmatrix} a_{11} & a_{12} & a_{13} \\ a_{21} & a_{22} & a_{23} \end{bmatrix}, \qquad (4)$$

taken in the order BA. The principle dictating the choice of order is this: a transformation is an *operation* or *operator*, acting on certain variables: the symbol of operation will always immediately precede the variables it affects, the operand. Thus when the variables x are transformed by A, the resulting variables are naturally symbolized by Ax; and when these new variables are in turn transformed by B, it is natural to write $B(Ax)$ for the outcome, and therefore natural to write BA for the matrix resulting from the two transformations in order.

By inspection we infer the general rule of

Matrix Multiplication. The element in the ith row and the jth column of the product matrix BA is obtained by multiplying the elements in the ith row of B into the corresponding elements in the jth column of A, and summing the products so obtained. In symbols, if B is of order $m \times n$ and A is of order $n \times p$ then $BA = C$, where C is of order $m \times p$, and

$$c_{ij} = \sum_{k=1}^{n} b_{ik} a_{kj}. \qquad \qquad (5)$$

It is important to note that multiplication is possible only if the number of columns in B is the same as the number of rows in A.

The reader should pause here to satisfy himself that he has grasped this fundamental rule. He should write out, in their literal fullness, a few products BA and AB for small values of m, n and p, and should construct and work out for himself some numerical examples also, such as

$$\begin{bmatrix} 1 & -2 & 3 \\ -4 & 2 & 5 \end{bmatrix} \begin{bmatrix} 1 & 3 \\ -1 & 0 \\ 2 & 4 \end{bmatrix} = \begin{bmatrix} 9 & 15 \\ 4 & 8 \end{bmatrix}.$$

It is to be observed that the typical element c_{ij} of BA is itself a matrix product ; in fact (5) is the same as

$$c_{ij} = [b_{i1} \; b_{i2} \ldots b_{in}] \; \{a_{1j} \; a_{2j} \ldots a_{nj}\}, \qquad . \qquad . \qquad (6)$$

where the first matrix on the right is a row vector, namely, the ith row of B, let us say $b'_{(i}$, and the second matrix is a column vector, the jth column of A, let us say $a_{(j}$. Thus $BA = [b'_{(i}a_{(j}]$.

EXAMPLES

(These will be not so much exercises as interspersed illustrations. For reference they will be numbered consecutively within each section, and will be distinguished by smaller print, without further title.)

1. The ith row of BA is the vector-matrix product $b'_{(i}A$.

2. The jth column of BA is $Ba_{(j}$.

We now come upon an important distinction between matrix algebra and ordinary scalar algebra ; namely, matrix multiplication is *non-commutative*. Consider for example

$$AB = \begin{bmatrix} a_{11} & a_{12} \\ a_{21} & a_{22} \end{bmatrix} \begin{bmatrix} b_{11} & b_{12} \\ b_{21} & b_{22} \end{bmatrix} = \begin{bmatrix} a_{11}b_{11} + a_{12}b_{21} & a_{11}b_{12} + a_{12}b_{22} \\ a_{21}b_{11} + a_{22}b_{21} & a_{21}b_{12} + a_{22}b_{22} \end{bmatrix}, \quad (7)$$

$$BA = \begin{bmatrix} b_{11} & b_{12} \\ b_{21} & b_{22} \end{bmatrix} \begin{bmatrix} a_{11} & a_{12} \\ a_{21} & a_{22} \end{bmatrix} = \begin{bmatrix} b_{11}a_{11} + b_{12}a_{21} & b_{11}a_{12} + b_{12}a_{22} \\ b_{21}a_{11} + b_{22}a_{21} & b_{21}a_{12} + b_{22}a_{22} \end{bmatrix}. \quad (8)$$

Scrutiny discloses, in this as in the general case, that *every* element in BA differs in form from the corresponding element in AB. Thus in general $BA \neq AB$. Indeed if A and B are not square but rectangular, then the only case in which BA and AB can coexist is when A is of order $m \times n$ and B is of order $n \times m$; and in such a case AB and BA are necessarily different, since the former is of order $m \times m$ while the latter is of order $n \times n$.

We must therefore always distinguish between *pre-multiplication* and *postmultiplication* of matrices ; in BA for example A is premultiplied by B, but in AB it is post-multiplied by B.

3. Since multiplication is non-commutative, the expanded form of $(A+B)^2$, defined as $(A+B)(A+B)$, is $A^2+AB+BA+B^2$, where A^2 is defined as $A(A)$.

4. Expand $(A+B)(A-B)$ and $(A-B)(A+B)$. Note that the expansions are different, and are of four, not of two terms.

7. Product of Three or More Matrices

Though not commutative, multiplication, when extended step by step to three or more factor matrices, is *associative*. For let C, B, A be matrices of order $m \times n$, $n \times p$, $p \times q$ respectively, and let us form $C(BA)$, where the brackets indicate that BA is first formed and is then premultiplied by C. We shall show that $C(BA)$ is identical with $(CB)A$. For since the $(h, j)^{th}$ element in BA is

$$\sum_{k=1}^{p} b_{hk}a_{kj}, \qquad \cdot \qquad \cdot \qquad \cdot \qquad (1)$$

the $(i, j)^{th}$ element in $C(BA)$ is, by the law of multiplication,

$$\sum_{h=1}^{n} c_{ih} \sum_{k=1}^{p} b_{hk}a_{kj}. \qquad \cdot \qquad \cdot \qquad (2)$$

But the double summation over np terms indicated here can be equally well carried out in the order indicated by

$$\sum_{k=1}^{p} \left(\sum_{h=1}^{n} c_{ih}b_{hk}\right)a_{kj}, \qquad \cdot \qquad \cdot \qquad (3)$$

which yields, though by a different order of operations, exactly the same final aggregate of np terms. Indeed it is customary to denote both of these equivalent double summations indiscriminately by

$$\sum_{h=1}^{n} \sum_{k=1}^{p} c_{ih} b_{hk} a_{kj}. \qquad . \qquad . \qquad . \quad (4)$$

The matrices $C(BA)$ and $(CB)A$ are thus element for element identical, and may therefore each be denoted without ambiguity by CBA. The reasoning may be extended step by step to the case of the product of any number of matrices. We conclude that matrix multiplication is associative.

1. Deduce from **6**, Ex. 1, that the $(i, j)^{th}$ element of UAX is $u_{(i}Ax_{(j}$, where $u_{(i}$ is the ith row of U, and $x_{(j}$ is the jth column of x. Deduce also that the ith row of UAX is $u_{(i}AX$ and that the jth column is $UAx_{(j}$.

2. In virtue of the associative nature of matrix multiplication the powers A^2, A^3, A^4, . . . of a matrix A (necessarily square) are defined without ambiguity, and identities such as $A^2(A) = A(A^2) = A^3$ and the like are valid.

3. Write down a few matrices of orders 2×2 and 3×3 with numerical elements. Find first their squares, then their cubes, checking the latter by using the two processes $A(A^2)$ and $A^2(A)$.

The Unit Matrix. The transformation illustrated here by the case $n = 3$,

$$\begin{aligned} y_1 &= x_1, \\ y_2 &= \quad x_2, \\ y_3 &= \qquad x_3, \end{aligned} \qquad . \qquad . \qquad . \qquad . \quad (5)$$

leaves the variables unchanged except in name. It is called the *identical transformation*, and the matrix corresponding to it,

$$\begin{bmatrix} 1 & . & . \\ . & 1 & . \\ . & . & 1 \end{bmatrix}, \qquad . \qquad . \qquad . \quad (6)$$

is called the *unit matrix*. (Here and elsewhere we shall use full stops to denote zero elements.) In the general case the unit matrix is a square matrix, of order $n \times n$, with unit elements in the *principal diagonal* (that is, the diagonal running from top left corner to the bottom right corner) and zero elements everywhere else; or, more briefly, such that $a_{ii} = 1$, $a_{ij} = 0$, $i \neq j$. It is denoted by I.

The reader is asked to verify for himself that $I^2 = I$, whence $I^3 = I$, and so on; also that $IA = AI = A$ for any matrix A, though it calls for remark that if A be rectangular, of order $m \times n$, then the premultiplying I in the above relation is of order $m \times m$, while the postmultiplying I is of order $n \times n$. With this proviso we can introduce or suppress an I at pleasure anywhere among the factors of a product matrix, as convenience may dictate.

Scalar Matrix. A matrix of the form λI, as for example in the case $n = 3$

$$\lambda I = \begin{bmatrix} \lambda & \cdot & \cdot \\ \cdot & \lambda & \cdot \\ \cdot & \cdot & \lambda \end{bmatrix}, \qquad . \qquad . \qquad . \quad (7)$$

is called a *scalar* matrix. Scalar multiplication, as defined in **5**, is equivalent to matrix multiplication (pre- or post-) by a scalar matrix. In symbols,

$$\lambda A = \lambda I A = A \lambda I = A \lambda. \qquad . \qquad . \quad (8)$$

Diagonal or Quasi–Scalar Matrix. A square matrix with its non-diagonal elements zero is called a *diagonal*, by some writers a *quasi-scalar*, matrix. It is thus defined by $a_{ij} = 0$, $i \neq j$.

4.

$$\begin{bmatrix} \lambda_1 & \cdot & \cdot \\ \cdot & \lambda_2 & \cdot \\ \cdot & \cdot & \lambda_3 \end{bmatrix}, \quad \begin{bmatrix} a_{11} & \cdot & \cdot & \cdot \\ \cdot & a_{22} & \cdot & \cdot \\ \cdot & \cdot & a_{33} & \cdot \\ \cdot & \cdot & \cdot & a_{44} \end{bmatrix},$$

are diagonal matrices.

5. The unit matrix I and the zero matrix 0 of order $n \times n$ are diagonal matrices.

$$\begin{bmatrix} I & \cdot & \cdot \\ \cdot & I & \cdot \\ \cdot & \cdot & I \end{bmatrix} \qquad , \qquad \begin{bmatrix} \cdot & \cdot & \cdot \\ \cdot & \cdot & \cdot \\ \cdot & \cdot & \cdot \end{bmatrix}$$

6. If A is a diagonal matrix of order $m \times m$ and B is any matrix of order $m \times n$, prove that AB is obtained from B by multiplying the *rows* of B respectively by a_{11}, a_{22}, ..., a_{nn}. In particular if x is a column vector,

$$Ax = \{a_{11}x_1 \ a_{22}x_2 \ \ldots \ a_{mm}x_m\}.$$

7. If A is a diagonal matrix of order $n \times n$, prove that BA is obtained from B by multiplying the *columns* of B respectively by a_{11}, a_{22}, . . ., a_{nn}. In particular, if u is a row vector,

$$uA = [u_1a_{11} \ u_2a_{22} \ \ldots \ u_na_{nn}].$$

8. Hence if A is a diagonal matrix, AB is not in general equal to BA. In order to have $AB = BA$ in general, B must be square and A must be not merely diagonal but scalar.

9. *Diagonal matrices of the same order are commutative in multiplication with each other.*

10. If the diagonal elements of a diagonal matrix D are d_i, prove that the matrix $c_0I + c_1D + c_2D^2 + \ldots + c_kD^k$ is diagonal, and that its diagonal elements are $c_0 + c_1d_i + c_2d_i^2 + \ldots + c_kd_i^k$.

8. Transposition : Interchange of Rows with Columns

The code has now been described, and the formal rules of operation are almost complete. It has appeared that matrices, under suitable restrictions upon their orders, obey the following laws,

$$A+B = B+A, \qquad A+(B+C) = (A+B)+C, \tag{1}$$
$$A(B+C) = AB+AC, \qquad (B+C)A = BA+CA,$$

which are the ordinary laws of elementary algebra, except for the omission of the commutative law of multiplication. Hence, with due vigilance regarding the order of factors in product terms, we may manipulate matrices in addition, subtraction and multiplication by the familiar rules.

1. Expand in full $(A+B)^3$ and $(A-B)^3$. Note that the expansions are of eight terms, which must be left separate.

2. Expand in full $(A-B)^2(A+B)$, $(A+B)(A-B)^2$.

3. Write down a number of simple products like the above, expand them in full and contrast the results with those of ordinary algebra, where the commutative law of multiplication is valid.

Transposition. There remains, however, an operation which has no analogue in ordinary algebra. From a matrix A we may construct a new matrix the rows of which are the columns of A, and consequently the columns of which are the rows of A. This operation is called *transposition*, and the resulting matrix is called the *transpose* of A and is denoted by A'. (In the less recent literature the word *conjugate* is used, but this word is already heavily overtaxed in other domains of mathematics.) In symbols, $A' \equiv [a_{ij}]' = [a_{ji}]$.

4. A matrix of order 1×1, that is, a scalar number, is unaltered by transposition.

5. The transpose of the column vector

$$x = \{x_1 \ x_2 ... x_n\}$$

is the row vector

$$x' = [x_1 \ x_2 ... x_n].$$

6.
$$\text{If } A = \begin{bmatrix} a_1 \ b_1 \ c_1 \\ a_2 \ b_2 \ c_2 \end{bmatrix}, \text{ then } A' = \begin{bmatrix} a_1 \ a_2 \\ b_1 \ b_2 \\ c_1 \ c_2 \end{bmatrix}.$$

7. Form the products AA' and $A'A$, where A is the matrix of Ex. 6, and remark on the nature of their diagonal elements. Consider the case of a general matrix A.

8. If $C = AA'$, observe that $c_{ij} = c_{ji}$, or $C = C'$.

9. If x is a column vector, form $x'x$ and observe that it is an ordinary number, a matrix of order 1×1, and that it is the sum of squares of the elements of x.

10. Examine in the same way xx', and show that it is a square matrix possessing the property $C' = C$.

11. Two operations of transposition restore the original matrix ; in symbols $(A')' = A$. The operation is thus *reflexive*.

Symmetric and Skew Symmetric Matrices. This new operation enables us to give simple and compact definitions of important special types of matrices. For example, if a matrix A is unaltered by transposition, it must be square and *symmetrical* about its principal diagonal, so that $a_{ji} = a_{ij}$. We therefore characterize a symmetric (or *axisymmetric*) matrix by $A' = A$.

If $A' = -A$, so that $a_{ji} = -a_{ij}$, the matrix A is called *skew symmetric* or *antisymmetric*. A skew symmetric matrix must necessarily have zero elements in the principal diagonal, since $a_{ii} = -a_{ii}$.

12. $A = \begin{bmatrix} a & h & g \\ h & b & f \\ g & f & c \end{bmatrix}$ is symmetric. $A' = \begin{bmatrix} a & h & g \\ h & b & f \\ g & f & c \end{bmatrix}$

13. $A = \begin{bmatrix} 0 & a & b \\ -a & 0 & c \\ -b & -c & 0 \end{bmatrix}$ is skew symmetric. $A' = \begin{bmatrix} 0 & -a & -b \\ +a & 0 & -c \\ +b & +c & 0 \end{bmatrix}$

14. A symmetric matrix of order $n \times n$ has in general $n(n+1)/2$ distinct elements.

15. A skew symmetric matrix of order $n \times n$ has in general $n(n-1)/2$ distinct elements.

9. The Transpose of a Product : Reversal Rule

Matrix multiplication is row-into-column ; in BA the rows of B are multiplied, element for element, into the columns of A. This being so, the transpose of BA cannot be $B'A'$, for that would imply columns of B multiplied into rows of A. It is in fact $A'B'$; for, writing $a_{ji} = a'_{ij}$, $b_{ji} = b'_{ij}$, we have

$$(BA)' = [\underset{k}{\Sigma} b_{jk} a_{ki}] = [\underset{k}{\Sigma} a'_{ik} b'_{kj}] = A'B'.$$

The reader, before going further, is asked to satisfy himself concerning this by constructing and inspecting actual examples, for this *reversal rule* in the transposition of products is of the greatest importance. We deduce from it step by step the general reversal rule :

$$(CBA)' = (BA)'C' \quad = A'B'C',$$
$$(DCBA)' = (CBA)'D' = A'B'C'D', \text{ and so on.}$$

The Complex Conjugate of a Matrix. If the elements of A are complex numbers a_{ij}, we denote their complex conjugates by $\bar{a}_{ij}$. The matrix $[\bar{a}_{ij}]$ is then called the *complex conjugate* of A, and is denoted by $\bar{A}$.

1. $(\bar{\bar{A}}) = A, (\overline{BA}) = \bar{B}\bar{A}$. Also $(\bar{A})' = (\bar{A}')$, denoted by $\bar{A}'$. or A^*

2. If a, b, c, d, e, f, g, h are all real numbers, and $i = \sqrt{-1}$, and if

$$A = \begin{bmatrix} a+ib & c+id \\ e+if & g+ih \end{bmatrix}, \text{ then } \bar{A}' = \begin{bmatrix} a-ib & e-if \\ c-id & g-ih \end{bmatrix}.$$

Hermitian and Skew Hermitian Matrices. If $\bar{A}' = A$ the matrix A is square and is unchanged by the operations of transposition and taking complex conjugates. Such a matrix is called *Hermitian*. On the other hand, if $\bar{A}' = -A$ the matrix A, which is necessarily square, is called *skew Hermitian*, or *anti-Hermitian*.

3. $$\begin{bmatrix} a & b+ic & e+if \\ b-ic & d & h+ik \\ e-if & h-ik & g \end{bmatrix} \text{ and } \begin{bmatrix} ia & -b+ic & -e+if \\ b+ic & id & -h+ik \\ e+if & h+ik & ig \end{bmatrix}$$

are respectively Hermitian and skew Hermitian ($i = \sqrt{-1}$).

4. *A Hermitian matrix which is real, that is, has exclusively real elements, is a real symmetric matrix. A skew Hermitian matrix with real elements is a real skew symmetric matrix.* Hence all theorems on Hermitian matrices include theorems on symmetric matrices as special cases.

5. If A is Hermitian, then iA is skew Hermitian. If A is skew Hermitian, then iA is Hermitian.

6. If A is Hermitian, it can be written as $R+iS$, where R is a real symmetric and S is a real skew symmetric matrix.

7. If A is a square matrix prove, by transposing, that $A'+A$ is symmetric and that $A'-A$ is skew symmetric. Also that $\bar{A}'+A$ is Hermitian, $\bar{A}'-A$ skew Hermitian.

8. Hence any square matrix A can be expressed as the sum of a symmetric and a skew symmetric matrix, since $A = \frac{1}{2}(A+A')+\frac{1}{2}(A-A')$; or as the sum of a Hermitian and a skew Hermitian matrix, since $A = \frac{1}{2}(A+\bar{A}')+\frac{1}{2}(A-\bar{A}')$.

9. *If A is any matrix, not necessarily square, then $A'A$ and AA' are both symmetric;* for by the reversal rule $(A'A)' = A'(A')' = A'A$, and similarly for AA'.

10. Prove in similar fashion that if A *is any matrix with complex elements then $\bar{A}'A$ and $A\bar{A}'$ are both Hermitian.*

11. Prove that matrices of the form $\begin{bmatrix} a & b \\ -b & a \end{bmatrix}$, $\begin{bmatrix} c & d \\ -d & c \end{bmatrix}$ are commutative, and that their laws of addition and multiplication are congruent with those of ordinary complex numbers $a+ib$, $c+id$, where $i = \sqrt{-1}$. Here we have a special matrix algebra in perfect correspondence, or isomorphism, with a familiar algebra.

10. Algebraic Expressions and Relations in Matrix Notation

The rules of matrix algebra being now established, we are able to write many commonly occurring expressions and relations with extreme conciseness. The following examples may serve as an indication. Those which introduce the matrix notation for *bilinear*, *quadratic* and *Hermitian forms* deserve close attention.

1. The sum of products, $a_1b_1 + a_2b_2 + \ldots + a_nb_n$, which is the typical element of a product matrix and which occurs also in questions of perpendicularity in coordinate geometry, in the differential of a function of several variables, and in a great variety of other situations in mathematics, may be written as a matrix of order 1×1, namely $a'b$, a product of a row vector a' and a column vector b, or equally well $b'a$. (The name *inner product* of vectors a and b is often used for the above sum, but the notation $a'b$ is precise, and calls for no special adjective.)

2. The sum of squares $x_1^2 + x_2^2 + \ldots + x_n^2$ is therefore $x'x$. For example the square of the distance, in Euclidean space, from the origin to a point having rectangular Cartesian coordinates $\{x_1\ x_2 \ldots x_n\}$ is $x'x$. The positive value of the square root of $x'x$ is called the *norm* of the vector x.

3. The reader will verify for himself that an expression such as $a_{11}x_1^2 + a_{22}x_2^2 + \ldots + a_{nn}x_n^2$ is the same as $x'Ax$, where

A is the diagonal matrix having elements a_{ii}. Also that $a_{11}x_1y_1 + a_{22}x_2y_2 + \ldots + a_{nn}x_ny_n$ is $x'Ay$, or $y'Ax$. See also Ex. 5 below.

4. *A set of equations of linear transformation may be written as the equality of two column vectors,* thus :

$$\begin{bmatrix} a_{11}x_1 + a_{12}x_2 + \ldots + a_{1n}x_n \\ a_{21}x_1 + a_{22}x_2 + \ldots + a_{2n}x_n \\ \cdots\cdots\cdots\cdots\cdots \\ a_{m1}x_1 + a_{m2}x_2 + \ldots + a_{mn}x_n \end{bmatrix} = \begin{bmatrix} y_1 \\ y_2 \\ \vdots \\ y_m \end{bmatrix} . \qquad (1)$$

Now the column vector on the right is simply y, and the column vector on the left is Ax, where A is of order $m \times n$ and x is of order $n \times 1$. Hence the set of equations is $Ax = y$. In the same way a set of simultaneous equations can be written $Ax = h$, where h is the column vector of constants on the right of the equations. From now on we adopt such abbreviated notations.

5. An expression such as

$$\underset{ij}{\Sigma\Sigma}a_{ij}x_iy_j, \qquad \begin{aligned} i &= 1, 2, \ldots, m, \\ j &= 1, 2, \ldots, n, \end{aligned}$$

of the first degree in each of two sets of variables x_i and y_j, is called a *bilinear form* in those variables. We may write it in the shape

$$x_1(a_{11}y_1 + a_{12}y_2 + \ldots + a_{1n}y_n) + x_2(a_{21}y_1 + a_{22}y_2 + \ldots + a_{2n}y_n)$$
$$+ \ldots + x_m(a_{m1}y_1 + a_{m2}y_2 + \ldots + a_{mn}y_n). \qquad (2)$$

But this is just $x'Ay$ written *in extenso*, since the bracketed expressions are the respective elements of the column vector Ay. This, then, is the matrix notation for a bilinear form ; and A is said to be the matrix of the form. Further, since a matrix of order 1×1 is unaltered by transposition, we may transpose the product $x'Ay$ and write the bilinear form alternatively as $y'A'x$. If $A = I$ the bilinear form is $x'Iy = x'y$, or $y'x$.

6. If the above bilinear form be denoted by ϕ, the partial derivatives $\partial\phi/\partial x_1$, $\partial\phi/\partial x_2$, $\ldots$, $\partial\phi/\partial x_m$ are the respective

bracketed expressions in (2), and these are the elements of the column vector Ay. Thus we may write

$$\partial(x'Ay)/\partial x' = Ay. \qquad . \qquad . \qquad . \qquad (3)$$

Here $\partial/\partial x'$ denotes an operation which may suitably be called *vector differentiation*; its formal resemblance to ordinary scalar differentiation is evident. In the same way we may write

$$\partial(x'Ay)/\partial y = x'A, \qquad . \qquad . \qquad . \qquad (4)$$

the vector of partial derivatives being now a row vector.

7. In the bilinear form $x'Ay$, let $x = y$ and let A be symmetric, $A' = A$. The bilinear form then becomes

$$\begin{aligned}
a_{11}x_1^2 &+ 2a_{12}x_1x_2 + 2a_{13}x_1x_3 + \ldots + 2a_{1n}x_1x_n \\
&+ a_{22}x_2^2 \quad + 2a_{23}x_2x_3 + \ldots + 2a_{2n}x_2x_n \\
&+ \ldots \ldots \ldots \ldots \\
&\qquad\qquad\qquad + a_{nn}x_n^2,
\end{aligned} \qquad (5)$$

or $\qquad \sum_{i\,j} \sum a_{ij}x_ix_j$, where $a_{ij} = a_{ji}$.

Such a form is called a *quadratic form*, of matrix A. Thus the matrix notation for a quadratic form is $x'Ax$, with the understanding that $A' = A$. If $A = I$, the quadratic form is $x'Ix = x'x$.

8. The equation of a central quadric with respect to the centre as origin, usually written

$$ax^2 + by^2 + cz^2 + 2fyz + 2gxz + 2hxy = 1, \qquad . \qquad (6)$$

will appear in matrix notation as $x'Ax = 1$. Written in full this would be

$$\begin{bmatrix} x & y & z \end{bmatrix} \begin{bmatrix} a & h & g \\ h & b & f \\ g & f & c \end{bmatrix} \begin{bmatrix} x \\ y \\ z \end{bmatrix} = 1, \qquad . \qquad . \qquad (7)$$

which reproduces almost exactly a descriptive notation used by certain writers of the last century.

9. The tangent plane to the quadric $x'Ax = 1$ at a point ξ (that is, column vector ξ of coordinates) upon it, or the polar plane with respect to a point ξ in the 3-dimensional Euclidean space, will be given by $x'A\xi = 1$ or equally well $\xi'Ax = 1$. The notation is the same for the analogue of this in any number of dimensions.

10. The reader will verify for himself that the row vector $\partial(x'Ax)/\partial x$ of partial derivatives of a quadratic form is $2x'A$, and that the column vector $\partial(x'Ax)/\partial x'$ is $2Ax$. Compare Ex. 6.

11. Let a set of variables or vector x be given linearly in terms of other variables u by the linear equations $x = Hu$, and in the same way let $y = Kv$. Then substituting in the bilinear form $x'Ay$ we obtain $u'H'AKv$, a new bilinear form of matrix $H'AK$. In the same way the transformation $x = Hu$ causes the quadratic form $x'Ax$ to become $u'H'AHu$, a quadratic form with matrix $H'AH$. The reader will easily show, by transposition, that $H'AH$ is symmetric.

12. A form $\bar{x}'Ax$, where A is Hermitian, so that $\bar{A}' = A$, is called a *Hermitian form*. Clearly if $x = Hu$ then $\bar{x}'Ax$ becomes $\bar{u}'\bar{H}'AHu$, and by transposing and taking the complex conjugate of $\bar{H}'AH$ we observe that $\bar{H}'AH$ is Hermitian. Three types of transformation of A, namely, $H'AK$, $H'AH$ and $\bar{H}'AH$, have thus been induced by the linear transformation of the variables occurring in bilinear, quadratic and Hermitian forms. These are called the *equivalent*, the *congruent* and the *conjunctive* transformations of A.

13. *A Hermitian form is real.* For by transposing and taking the complex conjugate we have

$$\overline{(\bar{x}'Ax)'} = \bar{x}'\bar{A}'(\overline{\bar{x}'})' = \bar{x}'Ax.$$

Thus the form, a matrix of order 1×1 and therefore a scalar number, is equal to its complex conjugate; hence it must be real.

11. Partitioned Matrices : Partitioned Multiplication

Submatrices. The array of elements belonging to (not necessarily consecutive) rows $i_1, i_2, \ldots, i_r$ and columns $j_1, j_2, \ldots, j_s$ constitutes a *submatrix* of A of order $r \times s$. In particular the elements of A are submatrices of order 1×1, the rows of A are submatrices of order $1 \times n$, and the columns of A are submatrices of order $m \times 1$.

It is a simple observation from the product rule

$$c_{ij} = \sum_k b_{ik} a_{kj}, \qquad \qquad \text{(1)}$$

where $C = BA$, that c_{ij} derives exclusively from the ith row of B and the jth column of A. It follows at once that the submatrix consisting of rows i_1, i_2, ..., i_r and columns j_1, j_2, ..., j_s of C will be the product of the submatrix consisting of these particular rows of B and the submatrix consisting of these particular columns of A. Thus *every submatrix in a product matrix is itself the product of two submatrices,* one from each of the factor matrices, in proper order.

If therefore we partition B by groups of rows into submatrices B_1, B_2, ..., B_i, ... in vertical array, and A by groups of columns into submatrices A_1, A_2, ..., A_j, ... in horizontal array, then the $(i, j)^{th}$ submatrix in BA, when BA is partitioned in row-groups exactly as B and in column-groups exactly as A, will be $B_i A_j$. The following is a simple illustration, the partitioning being indicated by dotted lines:

$$\begin{bmatrix} b_{11}\ b_{12} \\ b_{21}\ b_{22} \\ \hdashline b_{31}\ b_{32} \end{bmatrix} \begin{bmatrix} a_{11} & a_{12}\ a_{13} & a_{14} \\ a_{21} & a_{22}\ a_{23} & a_{24} \end{bmatrix} =$$

$$\begin{bmatrix} b_{11}a_{11}+b_{12}a_{21} & b_{11}a_{12}+b_{12}a_{22} & b_{11}a_{13}+b_{12}a_{23} & b_{11}a_{14}+b_{12}a_{24} \\ b_{21}a_{11}+b_{22}a_{21} & b_{21}a_{12}+b_{22}a_{22} & b_{21}a_{13}+b_{22}a_{23} & b_{21}a_{14}+b_{22}a_{24} \\ \hdashline b_{31}a_{11}+b_{32}a_{21} & b_{31}a_{12}+b_{32}a_{22} & b_{31}a_{13}+b_{32}a_{23} & b_{31}a_{14}+b_{32}a_{24} \end{bmatrix}. \quad \text{(2)}$$

We shall use a notation in which this result appears in the form

$$\begin{bmatrix} B_1 \\ B_2 \end{bmatrix} [A_1\ A_2\ A_3] = \begin{bmatrix} B_1A_1 & B_1A_2 & B_1A_3 \\ B_2A_1 & B_2A_2 & B_2A_3 \end{bmatrix}, \qquad \text{(3)}$$

submatrices being disposed like elements, the capital letters indicating that they are submatrices. The left side of (3) may also be written as $\{B_1\ B_2\} [A_1\ A_2\ A_3]$, the intention of the curled brackets being the same as in the notation of 4 (1) for column vectors. The analogy of (3) with the column-into-row product

$$\{b_1 \ b_2\} \ [a_1 \ a_2 \ a_3] = \begin{bmatrix} b_1 a_1 & b_1 a_2 & b_1 a_3 \\ b_2 a_1 & b_2 a_2 & b_2 a_3 \end{bmatrix} \qquad . \quad (4)$$

and the general result of like kind is evident.

A still more general law of multiplication of partitioned matrices is possible. In the result (1) we may break up the range of summation of k into sub-ranges, writing for example

$$\sum_{k=1}^{n} b_{ik} a_{kj} = \sum_{k=1}^{n_1} b_{ik} a_{kj} + \sum_{k=n_1+1}^{n_2} b_{ik} a_{kj} + \ldots + \sum_{k=n_h+1}^{n} b_{ik} a_{kj}, \quad . \quad (5)$$

or even using partial ranges which do not involve consecutive values of k; for example if $n = 6$ using 1, 3, then 2, 5, then 4, 6 for three partial ranges. Now k is both a column suffix of B and a row suffix of A, and so this division into partial ranges corresponds to partitioning the columns of B into groups in a certain way, and the rows of A into groups *in exactly the same way*. This will be called *conformable partitioning* of B and A. We may interpret each partial sum of terms on the right of (5) as the typical element of a product of two submatrices, the first being constituted by a certain group of columns of B, the second by the corresponding rows of A. Now combining this with the result typified by (3) we have such a result as

$$B = \begin{bmatrix} B_1 \\ B_2 \end{bmatrix} = \begin{bmatrix} B_{11} & B_{12} & B_{13} \\ B_{21} & B_{22} & B_{23} \end{bmatrix}, \ A = [A_1 \ A_2 \ A_3] = \begin{bmatrix} A_{11} & A_{12} & A_{13} \\ A_{21} & A_{22} & A_{23} \\ A_{31} & A_{32} & A_{33} \end{bmatrix},$$

$$BA = \begin{bmatrix} B_{11}A_{11} + B_{12}A_{21} + B_{13}A_{31} & B_{11}A_{12} + B_{12}A_{22} + B_{13}A_{32} \\ B_{21}A_{11} + B_{22}A_{21} + B_{23}A_{31} & B_{21}A_{12} + B_{22}A_{22} + B_{23}A_{32} \end{bmatrix} \qquad (6)$$
$$\begin{matrix} B_{11}A_{13} + B_{12}A_{23} + B_{13}A_{33}, \\ B_{21}A_{13} + B_{22}A_{23} + B_{23}A_{33}, \end{matrix}$$

where B and A have been conformably partitioned, and where the partitioning of the product BA agrees with the row-partitioning of B and the column-partitioning of A. The general result of this kind is proved in the same way, and its analogy with the ordinary rule of multiplication of matrices according to elements cannot fail to be observed. Formally we may enunciate it thus :

Theorem of Multiplication of Partitioned Matrices. Let B and A be two conformably partitioned matrices. Further, let the rows of B be partitioned arbitrarily into groups, and the columns of A also arbitrarily. Denote the submatrices of B and A so arising by B_{ij}, A_{ij}, where i denotes the order of occurrence in row groups, j in column groups. Finally let the product BA, or C, be partitioned according to the row-partitioning of B and the column-partitioning of A. Then the $(i, j)^{th}$ submatrix of C is

$$C_{ij} = \sum_k B_{ik} A_{kj}. \qquad . \qquad . \qquad . \quad (7)$$

In the case where all the submatrices are elements of B and A this result becomes the ordinary multiplication theorem for matrices.

The reader is again desired not to proceed until he has grasped, by examples both literal and numerical, and by examining the steps of a general proof along the lines we have indicated, the full implication of the multiplication theorem of partitioned matrices.

1. Consider the partitioned product

$$BA = [B_1\ B_2...B_n]\ \{A_1\ A_2...A_n\}$$
$$= B_1A_1 + B_2A_2 + ... + B_nA_n,$$

where the submatrices of B are its n columns, and of A its n rows. Examine the products B_kA_k. Observe that they are all matrices of order $m \times p$, and that by superimposing them in matrix addition we have BA in its usual form.

2. In Ex. 1 we have regarded the matrix B as a row vector having column vectors for its elements, and A as a column vector having row vectors for its elements. But we may equally well partition B according to its rows, regarding it, as it were, as a column vector having row vectors for its

elements, while partitioning A by its columns. We then have the product

$$BA = \{B_1 \; B_2 \ldots B_m\} \, [A_1 A_2 \ldots A_p] = \begin{bmatrix} B_1 A_1 & B_1 A_2 \ldots B_1 A_p \\ B_2 A_1 & B_2 A_2 \ldots B_2 A_p \\ \cdot \cdot \cdot \cdot \cdot \cdot \cdot \cdot \cdot \cdot \cdot \\ B_m A_1 & B_m A_2 \ldots B_m A_p \end{bmatrix},$$

where for example

$$B_i A_j = [b_{i1} \; b_{i2} \ldots b_{in}] \, \{a_{1j} \; a_{2j} \ldots a_{nj}\}$$
$$= b_{i1} a_{1j} + b_{i2} a_{2j} + \ldots + b_{in} a_{nj}.$$

This way of partitioning draws attention once again to the fact that the elements of BA are row-into-column vector products. Exs. 1 and 2 of **6** now appear as simple examples of partitioning; also Ex. 1 of **7**.

The algebra of matrices still lacks one important operation. *Division*, or the operation that plays the part of division, has yet to be defined. We should like to discover matrices *reciprocal* to A, that is, matrices R and S such that $RA = I$, $AS = I$. If these can be found it will be possible to solve a set of equations $Ax = h$ by simply premultiplying by R, for this would give $RAx = Rh$ or $x = Rh$, since $RA = I$. The key to the definition of a reciprocal matrix is therefore to be sought in the solution of simultaneous equations, and is found in that form of the solution which makes use of determinants. This will be the subject of the next chapter.

3. Let J be defined as a matrix of order $n \times n$ in which the non-zero elements are units in the diagonal at right angles to the principal diagonal, that is, the *secondary diagonal*. Thus $a_{i,\,n-i+1} = 1$, $a_{ij} = 0$, $i+j \neq n+1$. For example, if $n = 3$

$$J = \begin{bmatrix} \cdot & \cdot & 1 \\ \cdot & 1 & \cdot \\ 1 & \cdot & \cdot \end{bmatrix}.$$

Write out in full some products JA, where A is a rectangular matrix, and describe in words the effect on A. Do the same with products of the type AJ. Note that $J' = J$, $J^2 = I$.

4. Examine and describe in words the effect on a rectangular matrix A of the operation J_1AJ_2, where J_1 and J_2 are of the type J.

5. Let U be defined as a matrix of order $n \times n$ in which all elements are zero except those in the diagonal of $n-1$ places immediately above the principal diagonal (that is, the *superdiagonal*), these being units. Thus $a_{i,\,i+1} = 1$, $a_{ij} = 0$, $j-i \neq 1$. For example, if $n = 3$

$$U = \begin{bmatrix} . & 1 & . \\ . & . & 1 \\ . & . & . \end{bmatrix}.$$

Examine some products of type UA, U^2A, U^3A, ... ; AU, AU^2, ... ; $U'A$, $(U')^2A$, ... ; AU', $A(U')^2$, ... ; and so on. Note the effect on A of these various operations.

6. Do the same with UAU, $U'AU$, UAU', $U'AU'$, where A is square ; and with the corresponding cases where A is rectangular and pre- and postmultiplying matrices are of different orders.

7. Examine the powers of U, proving that $U^n = 0$. Also evaluate UU', $U'U$.

8. Prove that polynomials in A, such as

$$a_0I + a_1A + \ldots + a_kA^k,$$

are commutative in multiplication.

9. The sum of all elements in a general matrix A is equal to uAx, where all the elements of u and x are units.

10. Prove that if A and B are such that AB and BA coexist, then AB and BA have the same sum of diagonal elements. Establish a similar property for ABC, BCA, CAB, and conjecture and prove a general theorem.

11. Let H be a matrix derived by permuting in any way the rows of the unit matrix I. Then H' will be derived from I by the same permutation upon columns. Form $H'H$ and HH'. Inspect examples and establish a general theorem.

CHAPTER II

DEFINITION AND PROPERTIES OF DETERMINANTS

12. The Solution of Simultaneous Equations

LET us consider the elementary solution of simultaneous equations in 1, 2 and 3 unknowns.

(i) The solution of $a_{11}x_1 = h_1$ . . . . (1)

is $x_1 = h_1/a_{11}$, provided that $a_{11} \neq 0$. . . (2)

(ii) The solution of $a_{11}x_1 + a_{12}x_2 = h_1$,

$$a_{21}x_1 + a_{22}x_2 = h_2, \qquad (3)$$

found by elimination first of x_2, then of x_1, is

$$x_1 = (h_1 a_{22} - h_2 a_{12})/(a_{11}a_{22} - a_{12}a_{21}),$$
$$x_2 = (h_2 a_{11} - h_1 a_{21})/(a_{11}a_{22} - a_{12}a_{21}), \qquad (4)$$

provided that the denominator $a_{11}a_{22} - a_{12}a_{21}$, which is the same for both x_1 and x_2, does not vanish.

The expressions in the numerator and denominator of both solutions arise from the familiar procedure of *cross-multiplication* to bring about elimination, for example

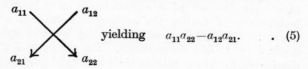

yielding $a_{11}a_{22} - a_{12}a_{21}$. . (5)

The general determinant, we shall find, is simply the extension of such cross-product formation as the above from a 2×2 matrix to an $n \times n$ matrix.

(iii) Solving in the same way the equations

$$a_{11}x_1 + a_{12}x_2 + a_{13}x_3 = h_1,$$
$$a_{21}x_1 + a_{22}x_2 + a_{23}x_3 = h_2, \qquad . \qquad . \quad (6)$$
$$a_{31}x_1 + a_{32}x_2 + a_{33}x_3 = h_3$$

by elimination of x_2 and x_3, we obtain x_1 in the form of a quotient the denominator of which is the six-termed expression

$$a_{11}a_{22}a_{33} - a_{11}a_{23}a_{32} + a_{12}a_{23}a_{31} - a_{12}a_{21}a_{33} + a_{13}a_{21}a_{32} - a_{13}a_{22}a_{31}, \quad (7)$$

the numerator being another six-termed expression obtained from the above by substituting h_1, h_2, h_3 respectively for a_{11}, a_{21}, a_{31}. Again x_2 and x_3 are quotients of similar nature with the same denominator as x_1, and this denominator must not vanish.

The solution of four simultaneous equations may be carried out in the same way by the various necessary eliminations; the denominator of the expressions for x_1, x_2, x_3, x_4 is then found to be the same in each case, namely an expression of 24 terms, 12 prefixed by positive and 12 by negative sign, each term being a product of four elements a_{ij}, no two of which in any term have the same first suffix or the same second suffix. The numerators are also found to be expressions of 24 terms.

It is natural to pause here, and instead of proceeding to the further heavy eliminations necessary for solving simultaneous equations in 5 or more unknowns, to study the properties of these numerators and denominators and thus by induction to define them generally, in the hope of solving the set of n equations in n unknowns and incidentally of finding the reciprocal matrix of a given matrix. The expressions constructed by this synthetic method are in fact *determinants*. The old name for them, *eliminants*, faithfully reflects their historical origin.

13. Salient Properties of Eliminants

The properties likely to be useful for purposes of definition emerge already in the examples we have given.

The respective denominators were a_{11}, $a_{11}a_{22} - a_{12}a_{21}$ and the longer expression (12, (7)) of 6 terms, presumably of $n!$ terms in general. For $n > 1$, apparently half of the terms are positive and half are negative. Each term consists of a product of n factors a_{ij}, and in these products all the first suffixes, 1, 2, ..., n and all the second suffixes 1, 2, ..., n appear once and once only. Hence if we order the factors in a term so that the first suffixes, which are really row suffixes of elements in the related square matrix A, are in natural order $123...n$, then the second suffixes will be some permutation $\alpha\beta\gamma...\nu$ of that natural order.

We are therefore led to define a determinant of order n as the following function of the elements a_{ij} of a square matrix A :

$$\Sigma \pm a_{1\alpha}a_{2\beta} \cdots a_{n\nu}, \qquad . \qquad . \qquad . \qquad (1)$$

the summation being extended over all the possible $n!$ permutations of natural order of second suffixes α, β, ..., ν.

It remains, however, to discover the rule by which the sign $+$ or $-$ is prefixed to any given term. This rule depends on the nature or *class* of the permutation $(\alpha\beta...\nu)$ of $(12...n)$; and this requires a study of the classification of permutations into even and odd classes.

14. Inversions, Transpositions, Even and Odd Permutations

When two indices in a permutation are out of natural order, the greater index preceding the lesser, like the 3 and the 2 in (1324), or the 5 and the 1 in (52341), such a derangement is called an *inversion*.

The interchange of two indices in a permutation without alteration of the rest, as when (13425) becomes (13524) by interchange of 4 with 5, is called a *transposition*.

The number of inversions of natural order in any permutation $(\alpha\beta...\nu)$ can be counted systematically and is unique. A permutation is said to be *even*, or of *even class*, when the number of inversions in it is even, and *odd*, or of *odd class*, when the number of inversions is odd.

15. Definition of Determinant

We are now able to define the *determinant* $|A|$ or $|a_{ij}|$ of a square matrix A of *order* $n \times n$ thus :

$$|A| \equiv |a_{ij}| \equiv \Sigma \pm a_{1a} a_{2\beta} \dots a_{n\nu}, \qquad . \qquad . \quad (1)$$

the summation, of $n!$ terms, being extended over all permutations $(\alpha\beta\dots\nu)$ of second or column suffixes of the elements a_{ij}, and the sign $+$ or $-$ being prefixed to any term according as the permutation is even or odd.

Permanent. The corresponding sum with terms all prefixed by the positive sign is called the *permanent* of A ; its properties are neither so simple nor so rich in application as those of determinants, but it has an importance in the theory of symmetric functions and in abstract algebra. We shall denote it by $\overset{+\ +}{|A|}$.

Notation of Determinants. The early writers, Cauchy, Jacobi and others, denoted $|A|$ as we have done above, by

$$\Sigma \pm a_{1a} a_{2\beta} \dots a_{n\nu}$$

or some similar notation. Cayley in 1841 introduced what has remained, for nearly a century, the standard notation

$$\begin{vmatrix} a_{11} & a_{12} \dots a_{1n} \\ a_{21} & a_{22} \dots a_{2n} \\ \cdot & \cdot \cdot \cdot \cdot \cdot \\ \cdot & \cdot \cdot \cdot \cdot \cdot \\ a_{n1} & a_{n2} \dots a_{nn} \end{vmatrix} \qquad . \qquad . \qquad . \qquad . \quad (2)$$

Other notations are $|a_{11} a_{22} \dots a_{nn}|$, the determinant being indicated by its diagonal elements ; or the single suffix notation $|a_1 b_2 \dots k_n|$, distinct letters being used to denote column order. Vandermonde, Sylvester and other writers used a so-called *umbral* notation, in which the letter a is discarded and the suffixes themselves are made predominant. In this notation

$$\begin{pmatrix} 1234 \\ 1234 \end{pmatrix}, \quad \begin{pmatrix} 2134 \\ 1234 \end{pmatrix}, \quad \begin{pmatrix} 2134 \\ 4231 \end{pmatrix} \qquad . \qquad . \quad (3)$$

would represent respectively the general determinant of the 4th order, the same with first and second rows interchanged, and the same with first and second rows and first and fourth columns interchanged.

The practice in the present book will be for the most part to use $|A|$ or $|a_{ij}|$, leaving the order to be understood from the context ; and we shall also indicate the determinant of a partitioned matrix by using the typical vertical bars of Cayley's notation ; thus,

$$\begin{vmatrix} A_1 & A_2 \\ \hline B_1 & B_2 \end{vmatrix} \quad \text{or} \quad \begin{vmatrix} A_1 & A_2 \\ B_1 & B_2 \end{vmatrix} . \qquad . \quad (4)$$

Other notations, in particular the one by diagonal elements (which however fails to distinguish between the determinant of A and that of A') will be used as occasion may require.

A few easy examples will serve to illustrate the classing of permutations by counting inversions, and the formation of determinants according to the definition.

1. To count the number of inversions in a permutation. Take for example (43521). The inversions are 43, 42, 41 ; 32, 31 ; 52, 51 ; and 21. In all, 8 inversions. The class of (43521) is therefore even, and the term $a_{14}a_{23}a_{35}a_{42}a_{51}$ in a determinant $|A| \equiv |a_{11}a_{22}a_{33}a_{44}a_{55}|$ of the 5th order would receive the positive sign.

2. The number of inversions may be counted by the aid of a diagram thus. Let the indices be written in an upper line, spaced out and in natural order, for example 1 2 3 4 5. Directly below them write in the same way the permutation in question, and then join each index in the upper line to the

same index in the lower line, curving the joins if necessary so that all intersections are of two joining lines and not more. For example :

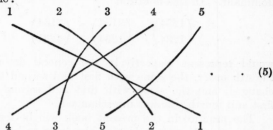

. . (5)

Here, since 4 and 3 are out of natural order in the lower line, the joins of 4 with 4 and 3 with 3 must intersect. The same is true for each inversion that is present in the lower line ; hence the number, as here eight, of intersections of the joining lines is equal to the number of inversions.

The reader is invited to try further examples. The diagram is merely a convenient device for giving visual form to the other method of counting inversions.

3. In the ways just indicated we may class the orders (123), (132), (231), (213), (312) and (321) as even and odd alternately. Hence the determinant $|A|$ of the 3rd order is

$$a_{11}a_{22}a_{33} - a_{11}a_{23}a_{32} + a_{12}a_{23}a_{31} - a_{12}a_{21}a_{33} + a_{13}a_{21}a_{32} - a_{13}a_{22}a_{31}, \quad . \ (6)$$

as has appeared already (**12**, (7)) in the denominator of the solution of three simultaneous equations.

Effect of Transposition. *A single transposition of order changes the class of a permutation.*

Consider $(a_1 a_2 a_3 a_4 a_5 a_6)$. Suppose a_2 and a_5 are interchanged. This may be done by passing a_2 over each of a_3, a_4, a_5 in turn, let us say in general over m indices, and then passing a_5 back over a_4, a_3, in general over $m-1$ indices. Now each passing over a single index either introduces an inversion or removes one, according as the two indices concerned were or were not previously in natural order ; and in either case the class must be changed. Hence we have in all an odd number, $2m-1$, of changes of class; so that a transposition changes the class of a permutation.

It follows that the difference between the number of inversions introduced and the number removed by a transposition must be an odd number.

Conjugate Permutations. Let us consider (3241) and (4213), and let us distinguish *index*, and *place occupied*, as follows. In (3241) the index 3 occupies the first place, place 1; in (4213) the index 1 occupies the third place, place 3. If the reader will examine the other indices in these two permutations, and the places they occupy in each, he will note that the numbers expressing index, and place occupied, in (3241) are exactly the numbers expressing place occupied, and index, in (4213). Two permutations in which the rôles played by index, and place occupied, are interchanged in this way are called *conjugate permutations*.

It is clear that a second interchange restores the original position; in other words the conjugate of a conjugate permutation is the original permutation. It is possible, too, for a permutation to be its own conjugate, or *self-conjugate*; the reader will verify, for example, that (13254) is self-conjugate.

4. Consider the diagram used in classing permutations. The upper line of numbers in natural order is simply what we have called place occupied; the lower line is what we have called index. Hence if we write the natural order such as 1 2 3 4 below the diagram of crossed lines and then complete the top line, the top line will give the conjugate permutation.

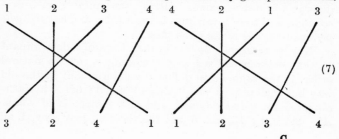

(7)

C

5. It follows at once that the diagram of a self-conjugate permutation must be *symmetrical* about its horizontal bisector.

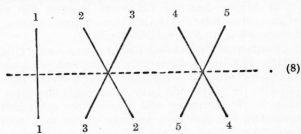

. (8)

6. If any term of a determinant such as $a_{1\alpha} \, a_{2\beta} \cdots a_{n\nu}$ has its factors reordered so that column suffixes $\alpha\beta\ldots\nu$ come into natural order, the term then becoming $a_{\alpha'1} \, a_{\beta'2} \cdots a_{\nu'n}$ then the resulting permutation of row suffixes $(\alpha'\beta'\ldots\nu')$ is the conjugate of the former permutation $(\alpha\beta\ldots\nu)$ of column suffixes.

The reason is that such a reordering is precisely the interchange of index with place occupied.

16. Identity of Class of Conjugate Permutations

There is a simple but important theorem, namely, that *conjugate permutations have the same number of inversions* of natural order.

Consider the diagrams used for counting inversions by the number of intersections, as exemplified in **15** (7). The diagrams are the same. Hence the number of intersections, and so of inversions, is the same ; and so *conjugate permutations are of the same class.*

Now to rearrange factors in each term of a determinant so that column suffixes, not row suffixes, are in natural order, is the same as to define the determinant in the old way but with respect to a matrix obtained from A by *interchanging rows and columns*, and this matrix is of course A'. Since, by **15**, Ex. 6 and the theorem just proved, all

terms in such an interchange retain the same sign as
before, we have the fundamental theorem :
The determinants of A and A' are the same.

EXAMPLE

Write out in full the six terms of the determinant $|A|$
of the 3rd order. Rearrange the factors·in each term so that
column suffixes are always in the order (123). Then put
$B = A'$, expand $|B|$ according to the original rule and compare
the results.

Relative Class of Two Permutations. To decide
whether two permutations of $(12...n)$ are of the same or
opposite class we need not count the inversions of each
with respect to $(12...n)$; it is sufficient to count their
relative inversions with respect to each other. For example
(3142) and (4213).

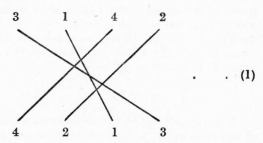

. . (1)

Here there are five intersections, so that five transpositions
of contiguous indices are required to remove the relative
inversions. (The moves are 3 with 1, 3 with 4, 3 with 2 ;
1 with 4, 1 with 2 ; these make 3142 become 4213.)
Hence the permutations are of opposite class.

It follows that the sign of a term in the expansion of a
determinant can be decided, not only by the number of
inversions of column suffixes when row suffixes are in
natural order, or by the number of inversions of row suffixes
when column suffixes are in natural order, but also by the

relative inversions of row with respect to column suffixes when the factors of the term are in *any* order.

Our example thus shows that the term $a_{34}a_{12}a_{41}a_{23}$, as likewise the term $a_{43}a_{21}a_{14}a_{32}$ in a determinant $|A|$ of the 4th order, would receive the negative sign.

Permutations have many other properties, but those we have just given are sufficient for developing various simple properties of determinants, which are of use in evaluating them by expansion in various ways.

17. Elementary Properties of Determinants

1. As we have seen, an immediate consequence of the fact that conjugate permutations have the same class is that $|A'| = |A|$.

2. Let columns j and k of A be interchanged. The natural order of row suffixes is not altered, but in all the terms of the determinant of the matrix B obtained in this way from A the column suffixes receive one transposition, namely, j with k. Hence, as compared with the same terms in $|A|$, the terms have undergone one change of sign. It follows that $|B| = -|A|$.

1.
$$\begin{vmatrix} 2 & -5 \\ 3 & 7 \end{vmatrix} = 29, \qquad \begin{vmatrix} -5 & 2 \\ 7 & 3 \end{vmatrix} = -29.$$

3. Since $|A'| = |A|$, it follows at once that interchange of two rows in $|A|$ produces a determinant $|B| = -|A|$.

2.
$$\begin{vmatrix} 2 & -5 \\ 3 & 7 \end{vmatrix} = 29, \qquad \begin{vmatrix} 3 & 7 \\ 2 & -5 \end{vmatrix} = -29.$$

4. If two rows, or two columns of A are identical, then $|A|$ is unaltered by their interchange; yet, as we have just seen, the new determinant is $-|A|$. Hence $|A| = -|A|$, so that $|A| = 0$. *Thus a determinant with two rows or two columns identical vanishes.*

5. For $n > 1$, a determinant $|A|$ possesses $n!/2$ positively signed terms and $n!/2$ negatively signed terms. (Determinantal sign due to permutation is meant, not the

adventitious sign that a term may have because a_{ij} happens, for instance, in one case to be 7 and in another to be -5.)

For let us put every $a_{ij} = 1$. Since all rows, or columns, are now identical we have $|A| = 0$. Also every term in the expansion of $|A|$ must be *numerically* equal to 1. Hence half of the terms are positive, half negative.

The following proof is also instructive. Of the $n!$ permutations, $n > 1$, suppose m are of even class. Subjecting these to a given transposition, such as the interchange of the index in the first place with that in the last, we obtain m permutations all different, as is readily seen, and all of odd class. Hence the number of permutations of odd class must be equal to or greater than the number of even class. By exactly similar reasoning the number of even class must be equal to or greater than the number of odd class. Hence even and odd permutations, for $n > 1$, are equinumerous.

6. The determinant of a diagonal matrix, since all non-diagonal elements are zero, reduces to a single term, namely, the product $a_{11}a_{22}...a_{nn}$ of the diagonal elements.

In particular the determinant of the unit matrix I is equal to 1, the determinant of a square zero matrix is equal to zero, and the determinant of λI is equal to λ^n.

7. If every a_{ij} be changed to λa_{ij}, every term in the determinant $|A|$ is multiplied by λ^n. Hence $|\lambda A| = \lambda^n |A|$. In particular $|-A| = (-)^n |A|$, $|-I| = (-1)^n$.

18. Primeness of a Determinant

A determinant, regarded as a multilinear polynomial in its elements a_{ij} (that is, linear in any one element), cannot be factorized into factors rational for *all* values of the a_{ij}. In other words it is a *prime polynomial* in its elements.

For let us assume on the contrary that it can be factorized into two polynomials θ, ϕ. Suppose that θ contains a certain element a_{ij}; then from the definition of $|A|$ it is clear that ϕ can contain no elements with row suffix i or column suffix j. Suppose again that ϕ contains a certain element a_{rs} ; then in the same way θ can contain no element with row suffix r or column suffix s. But this means that the elements a_{is} and a_{rj} can occur in neither factor, which is absurd, since they certainly occur in $|A|$. Hence the assumption is false, and $|A|$ is irreducible into polynomial factors.

19. Various Modes of Expansion of a Determinant

Expansion of $|A|$ by Elements of a Row or Column. What is the coefficient in $|A|$ of the leading element a_{11} ? Evidently it consists of all the terms

$$\Sigma \pm a_{2\beta}a_{3\gamma}\cdots a_{n\nu}, \qquad . \qquad . \qquad . \quad (1)$$

and since the row and column suffixes 1, 1 in a_{11} are in correct natural order they make no contribution to the sign of any term in (1). This sign is therefore given by the permutation $(\beta\gamma\ldots\nu)$ of the natural order $(23\ldots n)$.

For example, the coefficient of a_{11} in $|a_{11}a_{22}a_{33}|$ is $a_{22}a_{33}-a_{23}a_{32}$.

Now the aggregate of terms in (1) above, $(n-1)!$ in number, constitutes by definition the determinant of order $n-1$,

$$\begin{vmatrix} a_{22} & a_{23} & \cdots & a_{2n} \\ a_{32} & a_{33} & \cdots & a_{3n} \\ \cdots & \cdots & \cdots & \cdots \\ a_{n2} & a_{n3} & \cdots & a_{nn} \end{vmatrix} \qquad . \qquad . \qquad . \quad (2)$$

obtained by deleting from $|A|$ its first row and first column. We shall denote this determinant by $|A_{11}|$.

What is the coefficient of a_{ij} in the expansion of $|A|$? In $|A|$ let us pass the ith row upwards over the others until it becomes the first row, and then let us pass the jth column to the left over the others until it becomes the first column. The effect of this is to make a_{ij} become the leading element in a determinant derived from $|A|$ by $(i-1)+(j-1)$ changes of sign ; and the coefficient of a_{ij} in the expansion of this determinant is the determinant of order $n-1$ obtained by deleting the row and column containing a_{ij}. It follows that the coefficient or *cofactor* of a_{ij} in $|A|$ is the determinant obtained by suppressing the ith row and the jth column of $|A|$ and giving the sign $(-)^{i+j-2}$, that is, $(-)^{i+j}$ or equally well $(-)^{i-j}$ or $(-)^{j-i}$.

We shall denote this cofactor of a_{ij} in $|A|$ by $|A_{ij}|$. It could also be denoted by $\partial|A|/\partial a_{ij}$. The determinant obtained from $|A|$ by simply suppressing the ith row and j^{th} column, without giving sign $(-)^{i+j}$, is called a *minor* of $|A|$; and so we shall sometimes call a cofactor a *signed minor*. The determinant obtained by suppressing m rows and m columns of $|A|$ is called a minor of order $n-m$.

Now in the expression

$$a_{11}|A_{11}|+a_{12}|A_{12}|+...+a_{1n}|A_{1n}| \qquad . \qquad (3)$$

there are altogether $(n-1)!n$ terms, that is, $n!$ terms, all different and all occurring with correct sign in $|A|$. But the expansion of $|A|$ itself has only $n!$ terms. Hence the expansion (3) is a complete expansion of $|A|$. It will be called the expansion of $|A|$ according to elements of the first row and their cofactors.

In exactly the same way, since we can bring any row into first place with no more than a change of sign in $|A|$, we see that there is an expansion of $|A|$ according to elements of any row, namely,

$$|A| = a_{i1}|A_{i1}|+a_{i2}|A_{i2}|+...+a_{in}|A_{in}|, \qquad . \qquad (4)$$

and further, since $|A'| = |A|$, according to elements of any column and their cofactors.

The following examples give useful reductions and transformations of determinants, depending on the properties thus far derived.

1.
$$\begin{vmatrix} a_1 & b_1 & c_1 \\ a_2 & b_2 & c_2 \\ a_3 & b_3 & c_3 \end{vmatrix} = a_1 \begin{vmatrix} b_2 & c_2 \\ b_3 & c_3 \end{vmatrix} - b_1 \begin{vmatrix} a_2 & c_2 \\ a_3 & c_3 \end{vmatrix} + c_1 \begin{vmatrix} a_2 & b_2 \\ a_3 & b_3 \end{vmatrix}$$

$$= a_1 \begin{vmatrix} b_2 & c_2 \\ b_3 & c_3 \end{vmatrix} - a_2 \begin{vmatrix} b_1 & c_1 \\ b_3 & c_3 \end{vmatrix} + a_3 \begin{vmatrix} b_1 & c_1 \\ b_2 & c_2 \end{vmatrix},$$

and four other similar expansions.

2.
$$\begin{vmatrix} 1 & -3 & 7 \\ -2 & 4 & 5 \\ 3 & 1 & -1 \end{vmatrix} = 1 \begin{vmatrix} 4 & 5 \\ 1 & -1 \end{vmatrix} - (-3) \begin{vmatrix} -2 & 5 \\ 3 & -1 \end{vmatrix} + 7 \begin{vmatrix} -2 & 4 \\ 3 & 1 \end{vmatrix}$$

$$= -9 + 3 \, (-13) + 7 \, (-14)$$

$$= -146.$$

The reader should evaluate this determinant according to elements of rows and of columns and cofactors in all the other possible ways, and should invent and work through many other numerical examples of the same kind.

3.
$$\begin{vmatrix} a_1 & \cdot & \cdot & \cdot \\ a_2 & b_2 & c_2 & d_2 \\ a_3 & b_3 & c_3 & d_3 \\ a_4 & b_4 & c_4 & d_4 \end{vmatrix} = a_1 \begin{vmatrix} b_2 & c_2 & d_2 \\ b_3 & c_3 & d_3 \\ b_4 & c_4 & d_4 \end{vmatrix} \qquad . \qquad . \quad \textbf{(5)}$$

by expanding the determinant on the left in terms of its first row.

4.
$$\begin{vmatrix} a_1 + \lambda a_1 & b_1 + \lambda \beta_1 & c_1 + \lambda \gamma_1 \\ a_2 & b_2 & c_2 \\ a_3 & b_3 & c_3 \end{vmatrix} = \begin{vmatrix} a_1 & b_1 & c_1 \\ a_2 & b_2 & c_2 \\ a_3 & b_3 & c_3 \end{vmatrix} + \lambda \begin{vmatrix} a_1 & \beta_1 & \gamma_1 \\ a_2 & b_2 & c_2 \\ a_3 & b_3 & c_3 \end{vmatrix} \quad \textbf{(6)}$$

again by expanding in terms of the first row.

5. In the above, put $[a_1 \ \beta_1 \ \gamma_1] = [a_2 \ b_2 \ c_2]$ or $= [a_3 \ b_3 \ c_3]$. Then the second term on the right involves a determinant with two rows identical and so vanishes. Hence the important result :

A determinant is unaltered in value when to any row, or column, is added a constant multiple of any other row, or column.

6. Evaluate the determinant of Ex. 2 by using this fact.

One way is as follows. Perform on the determinant the operations $\text{row}_2 + 2\ \text{row}_1$, and then $\text{row}_3 - 3\ \text{row}_1$, the row to be modified being always written first in these operations. Then expand in terms of the first column. This gives

$$\begin{vmatrix} 1 & -3 & 7 \\ 0 & -2 & 19 \\ 0 & 10 & -22 \end{vmatrix} = \begin{vmatrix} -2 & 19 \\ 10 & -22 \end{vmatrix} = -146.$$

7. Perform on the same determinant of Ex. 2 the operations $\text{col}_2 + 3\ \text{col}_1$, $\text{col}_3 - 7\ \text{col}_1$, and then evaluate as before.

8. $\begin{vmatrix} \lambda a_1 & \lambda b_1 & \lambda c_1 \\ \mu a_2 & \mu b_2 & \mu c_2 \\ \nu a_3 & \nu b_3 & \nu c_3 \end{vmatrix} = \begin{vmatrix} \lambda a_1 & \mu b_1 & \nu c_1 \\ \lambda a_2 & \mu b_2 & \nu c_2 \\ \lambda a_3 & \mu b_3 & \nu c_3 \end{vmatrix} = \lambda \mu \nu \begin{vmatrix} a_1 & a_2 & a_3 \\ b_1 & b_2 & b_3 \\ c_1 & c_2 & c_3 \end{vmatrix}.$ (7)

9. By means of the results (i) $|\lambda A| = \lambda^n |A|$ and (ii) $|A'| = |A|$ prove that the determinant of a skew symmetric matrix S of odd order vanishes identically, and that the determinant of a skew Hermitian matrix of odd order is a purely imaginary complex number.

(In the first case $|S| = -|S|$, in the second $|S| = -|\bar{S}|$.)

10. $\begin{vmatrix} 1 & 1 & 1 \\ a & b & c \\ a^2 & b^2 & c^2 \end{vmatrix} = (c-b)(c-a)(b-a).$.   . (8)

(By any of several modes of expansion, the determinant is seen to be homogeneous in a, or b, or c. Further, if $b = a$, or $c = a$, or $b = c$, the determinant has two columns equal and so vanishes. Hence by the Remainder Theorem it contains $(b-a)$, $(c-a)$ and $(c-b)$ as factors. But the term arising from the principal diagonal, the leading term, is bc^2. Hence the remaining factor is purely numerical and is equal to unity.)

11. The determinant of Ex. 10 is a case of an *alternant*, so called because to interchange any two of a, b, c is to interchange two columns and therefore to alter the sign of the polynomial. If we write this alternant as $|a^0\ b^1\ c^2|$, the reader will prove similarly that the general alternant $|a^0 b^1 c^2 \ldots k^{n-1}|$ is equal to the continued product of all the differences that can be formed from the $\frac{1}{2}n(n-1)$ pairs of letters taken from

a, b, ..., k, the alphabetical order being reversed in each pair. The polynomial so derived is called the *difference-product* of a, b, ..., k and will be denoted by $\Delta(a, b, ..., k)$.

12. $|a^1 b^2 ... k^n| = ab...k\Delta(a, b, ... k) = \Delta(0, a, b, ..., k)$.

13. $|a_1 b_2 ... k_n|$ is the alternant $|a^1 b^2 ... k^n|$ with indices lowered into suffixes. In fact a determinant may be regarded as the result of a symbolic operation of difference-product type, in the sense that if E_a is an operator defined by $E_a a_i = a_{i+1}$, and E_b, ..., E_k are similar independent operators, then

$$\Delta(0, E_a, E_b, ..., E_k) a_0 b_0 ... k_0 = |a_1 b_2 ... k_n|.$$

14. If we write the general alternant as

$$| a_1^{\,0} \, a_2^{\,1} \, ... \, a_n^{\,n-1} |$$

we may note that the element in the ith row and j^{th} column is $a_j^{\,i-1}$. Hence an alternant matrix is briefly characterized by

$$\left[a_j^{\,i-1} \right], \text{ and its transpose by } \left[a_i^{\,j-1} \right].$$

15.
$$\begin{vmatrix} a_{11} & a_{12} & ... & a_{1n} \\ . & a_{22} & ... & a_{2n} \\ . & . & . & . \\ . & . & . & a_{nn} \end{vmatrix} = a_{11} a_{22} ... a_{nn}. \qquad . \qquad . \quad (9)$$

where $a_{ij} = 0$, $i > j$. (All elements below diagonal are zero.) To prove this, expand in terms of the first column, and repeat the operation on the determinants of lower order so obtained. Alternatively, note that any term involving elements from above the diagonal must necessarily also involve elements from below the diagonal, and these are all zero. Thus only the diagonal or leading term survives.

Length, Area, Volume and Hypervolume as Determinants. In the following examples we develop in determinantal form the length of a straight line segment, the area of a plane triangle, the volume of a tetrahedron and the hypervolumes of the successive analogues of these figures in space of higher dimensions.

16. The line segment. If A is on a straight axis OX at distance x_1 from the origin O, and B is at distance x_2, the signed length BA is $x_1 - x_2$, and this is equal to

$$\begin{vmatrix} x_1 & 1 \\ x_2 & 1 \end{vmatrix} \text{ or } \begin{vmatrix} x_1 - a & 1 \\ x_2 - a & 1 \end{vmatrix}, \qquad . \qquad . \qquad (10)$$

the equality of these expressing the fact that the length is invariant under change of origin from $x = 0$ to $x = a$. If B is the origin, $x_2 = 0$ and the length is merely x_1.

17. The triangle. Let the reader draw a triangle ABC in the first quadrant XOY of ordinary rectangular axes, and perpendiculars from A $= (x_1, y_1)$, B $= (x_2, y_2)$, C $= (x_3, y_3)$ to OX to meet OX in A', B', C'. Then, with due regard for sign, as given by direction of base segments, the area ABC is seen to be the sum of the areas of trapezia on bases B'A', A'C', C'B', the first trapezium for example being B'A'AB. Since the area of a trapezium is the length of the base multiplied by the mean of the extreme ordinates, the area ABC is by (10)

$$\tfrac{1}{2} \left\{ (y_1 + y_2) \begin{vmatrix} x_1 & 1 \\ x_2 & 1 \end{vmatrix} + (y_2 + y_3) \begin{vmatrix} x_2 & 1 \\ x_3 & 1 \end{vmatrix} + (y_3 + y_1) \begin{vmatrix} x_3 & 1 \\ x_1 & 1 \end{vmatrix} \right\}$$

$$= \tfrac{1}{2} \begin{vmatrix} x_1 & 1 & y_2 + y_3 \\ x_2 & 1 & y_3 + y_1 \\ x_3 & 1 & y_1 + y_2 \end{vmatrix} = \tfrac{1}{2} \begin{vmatrix} x_1 & y_1 & 1 \\ x_2 & y_2 & 1 \\ x_3 & y_3 & 1 \end{vmatrix}, \qquad . \qquad . \qquad . \qquad (11)$$

the last determinant being obtained from the one on the left of it by the operations $\text{col}_3 - (y_1 + y_2 + y_3) \, \text{col}_2$, followed by interchange of col_2 and col_3.

The determinant (11) is easily seen to be unaltered if each x_i is replaced by $x_i - a$, each y_i by $y_i - b$, the invariance of area under change of origin being thus expressed. Note also that if C is the origin $(x_3, y_3) = (0, 0)$ and the last row and column of the determinant in (11) may then be deleted.

18. The tetrahedron. With a similar naming of points, let the reader draw a tetrahedron in the first octant OXYZ, and perpendiculars from A, B, C, D to the plane XOY to meet it in A', B', C', D'. (In a first drawing he may let A' fall within the triangle B'C'D'. In later drawings he should consider varied orientations.) Then the volume of the tetrahedron is seen to be the sum of signed volumes of prisms on triangular bases $B'C'A'$, $C'A'D'$, $A'D'B'$, $D'B'C'$. Since the volume of a right prism is the base area multiplied by the mean of the vertical edges, the volume of ABCD is by (11)

$$\frac{1}{6}\begin{vmatrix} x_1 & y_1 & 1 & z_2+z_3+z_4 \\ x_2 & y_2 & 1 & z_3+z_4+z_1 \\ x_3 & y_3 & 1 & z_4+z_1+z_2 \\ x_4 & y_4 & 1 & z_1+z_2+z_3 \end{vmatrix} = \frac{1}{6}\begin{vmatrix} x_1 & y_1 & z_1 & 1 \\ x_2 & y_2 & z_2 & 1 \\ x_3 & y_3 & z_3 & 1 \\ x_4 & y_4 & z_4 & 1 \end{vmatrix}, \quad . \quad (12)$$

the last determinant being obtained from the one on the left of it by the operations $\mathrm{col}_4-(z_1+z_2+z_3+z_4)\,\mathrm{col}_3$, followed by interchange of col_3 and col_4.

The form of the general result is now plain. The determinant occurring in it is of order $n+1$ and of the type of (12), and the outer factor is $1/n!$. If the last corner is the origin, all elements of the last row except the final unit are zero, and the determinant reduces to one of the nth order, like that obtained by deleting the last row and column of (12). It is also clearly invariant under a change of origin from $(0, 0, ..., 0)$ to $(a, b, ..., k)$.

In the corresponding figures of completed type, the line segment, the parallelogram (sum of two equal triangles), the parallelehedron (sum of six equal tetrahedra) and so on, the length, area, volume and hypervolumes are expressed by the same determinant, but without the factor $1/n!$. These determinantal forms are valuable in the discussion of differential elements of volume in space of 2, 3 or more dimensions.

Elementary Operations on Determinants. The following examples are important as showing that certain commonly occurring elementary operations on determinants, such as interchanging or permuting their rows or columns, or adding to any row (or column) some multiple of any other row (or column), can be effected by pre- or postmultiplication of the matrix A of the determinant by suitable matrices.

19. Interchanging two rows. Consider

$$\begin{bmatrix} . & 1 & . \\ 1 & . & . \\ . & . & 1 \end{bmatrix}\begin{bmatrix} a_1 & b_1 & c_1 \\ a_2 & b_2 & c_2 \\ a_3 & b_3 & c_3 \end{bmatrix} = \begin{bmatrix} a_2 & b_2 & c_2 \\ a_1 & b_1 & c_1 \\ a_3 & b_3 & c_3 \end{bmatrix}. \quad . \quad (13)$$

Premultiplication of A by a matrix obtained by interchanging the first two rows of I has effected the interchange of the first two rows of A. The reader will prove that in general any permutation of the rows of a matrix A can be effected by premultiplying A by a matrix derived from I by the same permutation of rows. Compare 11, Ex. 3.

20. Any permutation of the columns of A can be effected by postmultiplying A by a matrix derived from I by the same permutation of columns. This is simply the result of Ex. 19 transposed.

21.
$$\begin{bmatrix} 1 & \lambda & . \\ . & 1 & . \\ . & . & 1 \end{bmatrix} \begin{bmatrix} a_1 & b_1 & c_1 \\ a_2 & b_2 & c_2 \\ a_3 & b_3 & c_3 \end{bmatrix} = \begin{bmatrix} a_1 + \lambda a_2 & b_1 + \lambda b_2 & c_1 + \lambda c_2 \\ a_2 & b_2 & c_2 \\ a_3 & b_3 & c_3 \end{bmatrix}. \quad (14)$$

Thus operations such as $\text{row}_i + \lambda \text{row}_k$ are effected by a premultiplication by I with a λ inserted in the $(i, k)^{th}$ position. The reader will at once see the suitable premultiplication for $\text{row}_i + \lambda \, \text{row}_h + \mu \, \text{row}_k$, and so on.

22. By transposing Ex. 21 we see the suitable postmultiplications that will effect $\text{col}_j + \lambda \, \text{col}_h + \mu \, \text{col}_k$ and the like.

23. Observe that
$$\begin{bmatrix} 1 & \lambda & . \\ . & 1 & . \\ . & . & 1 \end{bmatrix} \begin{bmatrix} 1 & . & . \\ . & 1 & \nu \\ . & . & 1 \end{bmatrix} \begin{bmatrix} 1 & . & \mu - \lambda \nu \\ . & 1 & . \\ . & . & 1 \end{bmatrix} = \begin{bmatrix} 1 & \lambda & \mu \\ . & 1 & \nu \\ . & . & 1 \end{bmatrix} . \quad (15)$$

and show in general that any matrix with units in the principal diagonal and all elements zero on one side of the diagonal can be expressed as a product of matrices representing elementary operations on rows or columns.

24. From expansion by a row, or from Ex. 8 above, if all elements in a row of $|A|$ are changed in sign the resulting determinant is equal to $-|A|$. A useful consequence is that if the 2nd, 4th, 6th, ... rows and the 2nd, 4th, 6th, ... columns of $|A|$ are changed in sign the determinant is unaltered in value. In a brief notation $|(-)^{i+j} a_{ij}| = |a_{ij}|$.

20. Practical Evaluation of Determinants by Condensation

The method of expanding a determinant according to elements of a row or column is in general of practical service only when the determinant is of no higher than the

3rd order. For higher orders the process has to be repeated, and the number of terms soon becomes very large. For this reason less cumbrous methods based on *condensation*, that is, on the systematic reduction of a determinant to one of lower order, are much superior in practice. Most of them are variants of a method ascribed to Chiò (1853), but virtually used by Gauss more than forty years earlier in evaluating symmetric determinants.

Let us consider the case of the 4th order,

$$|A| = |a_1 b_2 c_3 d_4| = \begin{vmatrix} a_1 & b_1 & c_1 & d_1 \\ a_2 & b_2 & c_2 & d_2 \\ a_3 & b_3 & c_3 & d_3 \\ a_4 & b_4 & c_4 & d_4 \end{vmatrix}, \qquad . \qquad . \quad (1)$$

and let us suppose meanwhile that $a_1 \neq 0$. No generality is lost, because if a_1 were zero we could bring some non-zero element to leading position by an interchange of rows and of columns, altering at most the sign of $|A|$. Now let us multiply all the rows of $|A|$ except the first by a_1. Finally, let us perform the operations

$$\text{row}_2 - a_2\text{row}_1, \ \text{row}_3 - a_3\text{row}_1, \ \text{row}_4 - a_4\text{row}_1.$$

The result is

$$a_1^3 \, |A| = \begin{vmatrix} a_1 & b_1 & c_1 & d_1 \\ a_1 a_2 & a_1 b_2 & a_1 c_2 & a_1 d_2 \\ a_1 a_3 & a_1 b_3 & a_1 c_3 & a_1 d_3 \\ a_1 a_4 & a_1 b_4 & a_1 c_4 & a_1 d_4 \end{vmatrix} = \begin{vmatrix} a_1 & b_1 & c_1 & d_1 \\ 0 & |a_1 b_2| & |a_1 c_2| & |a_1 d_2| \\ 0 & |a_1 b_3| & |a_1 c_3| & |a_1 d_3| \\ 0 & |a_1 b_4| & |a_1 c_4| & |a_1 d_4| \end{vmatrix}$$

$$= a_1 \begin{vmatrix} |a_1 b_2| & |a_1 c_2| & |a_1 d_2| \\ |a_1 b_3| & |a_1 c_3| & |a_1 d_3| \\ |a_1 b_4| & |a_1 c_4| & |a_1 d_4| \end{vmatrix}. \qquad . \qquad . \quad (2)$$

A rule of condensation is thus derived, which may be formulated in the general case as follows :

Let a_1 be called the *pivotal element* or *pivot*. Then the determinant $|A|$ of order n is equal to $1/a_1^{n-2}$ multiplied by a determinant of order $n-1$, the elements of which are the minors of $|A|$ of order 2 having a_1 as leading element, placed in proper priority of rows and columns. Thus the minor belonging to the 1st and $(i+1)^{th}$ rows, 1st and $(j+1)^{th}$ columns of $|A|$ will be the $(i, j)^{th}$ element of the condensed determinant.

1.　　　　$$|a_1 b_2 c_3| = a_1^{-1} \begin{vmatrix} |a_1 b_2| & |a_1 c_2| \\ |a_1 b_3| & |a_1 c_3| \end{vmatrix}.$$　.　　.　. **(3)**

2. If A is symmetric and the pivot is the leading element a_{11}, the condensed determinant is also symmetric.

The pivot need not be the leading element, which may be zero. Any convenient non-zero element a_{ij} in $|A|$ may be taken as pivot. The condensed determinant then has for elements all the minors of order 2 containing a_{ij}, each taken as if a_{ij} were its leading element and given sign accordingly. In other words, the diagonal of these minors which contains the pivot a_{ij}, no matter what orientation it has in $|A|$, is looked upon as the *principal diagonal*. Then $|A|$ is evaluated as before, except that the sign $(-)^{i+j}$ must be imposed.

3.　　$$\begin{vmatrix} -4 & 3 & 5 \\ 2 & -4 & -3 \\ 5 & -2 & -7 \end{vmatrix} = (-)^{1+2} \frac{1}{3^{3-2}} \begin{vmatrix} -10 & 11 \\ 7 & -11 \end{vmatrix}$$

$$= -\frac{1}{3}(33) = -11.$$

The pivot is italicized. The reader should work through for himself, checking the result by choosing other pivots, as well as by expanding in terms of a row or column. He should also do **19**, Ex. 2 by pivotal condensation.

A determinant of order n being reduced by a first pivotal condensation to one of order $n-1$, the latter in its turn can be reduced by a second pivotal condensation to one of order $n-2$, and so on until finally we have a determinant of order 2, a simple cross-product. Naturally the case where the pivot is 1 or -1 will give the least arithmetical work. It is always possible, by a preliminary division of the row or column containing any non-zero element, to turn that element into a unit pivot. This may introduce decimal fractions and approximations into the computation ; in practical applications, however, in which the elements are in any case approximate, rounded off to a certain number of digits, no disadvantage ensues from this.

4.
$$\begin{vmatrix} -2 & 3 & 2 & -5 \\ 3 & -4 & -5 & 6 \\ 4 & -7 & -6 & 9 \\ -3 & 5 & 4 & -10 \end{vmatrix} = -\frac{1}{3^2} \begin{vmatrix} 1 & -7 & -2 \\ -2 & -4 & -8 \\ 1 & 2 & -5 \end{vmatrix}.$$

The pivot is again italicized. The reader should complete the evaluation, and should repeat the work with different pivots. The value of the determinant is -18.

Leading pivots are the most advantageous. Next to them in convenience come diagonal pivots.

Some Important Identities. It may be observed that pivotal condensation corresponds exactly to the process of successive elimination by which simultaneous linear equations are solved. Thus if we have a set of equations $Ax = h$ from which we eliminate x_j, using the ith equation for this purpose, the matrix of the equations so derived is obtained by condensing A with respect to the pivot a_{ij}. The reader should verify this.

The method of condensation gives rise to a hierarchy of useful identities. We have in the first place

$$|a_1 b_2 c_3| = a_1^{-1} \begin{vmatrix} |a_1 b_2| & |a_1 c_2| \\ |a_1 b_3| & |a_1 c_3| \end{vmatrix}, \quad a_1 \neq 0. \qquad . \quad (4)$$

Next we have

$$|a_1 b_2 c_3 d_4| = a_1^{-2} \begin{vmatrix} |a_1 b_2| & |a_1 c_2| & |a_1 d_2| \\ |a_1 b_3| & |a_1 c_3| & |a_1 d_3| \\ |a_1 b_4| & |a_1 c_4| & |a_1 d_4| \end{vmatrix}, \quad a_1 \neq 0. \qquad (5)$$

Carrying out the condensation one stage further, with

leading pivot (assumed non-zero) $|a_1b_2|$, and using (4) in the evaluation of every minor, we derive at once

$$|a_1b_2c_3d_4| = |a_1b_2|^{-1} \begin{vmatrix} |a_1b_2c_3| & |a_1b_2d_3| \\ |a_1b_2c_4| & |a_1b_2d_4| \end{vmatrix} . \qquad (6)$$

In exactly the same way, at the second stage of condensation of $|a_1b_2c_3d_4e_5|$ we arrive at

$$|a_1b_2c_3d_4e_5| = |a_1b_2|^{-2} \begin{vmatrix} |a_1b_2c_3| & |a_1b_2d_3| & |a_1b_2e_3| \\ |a_1b_2c_4| & |a_1b_2d_4| & |a_1b_2e_4| \\ |a_1b_2c_5| & |a_1b_2d_5| & |a_1b_2e_5| \end{vmatrix} . \quad (7)$$

Now condensing once again, with leading pivot (assumed non-zero) $|a_1b_2c_3|$, and using (6) repeatedly, we obtain

$$|a_1b_2c_3d_4e_5| = |a_1b_2c_3|^{-1} \begin{vmatrix} |a_1b_2c_3d_4| & |a_1b_2c_3e_4| \\ |a_1b_2c_3d_5| & |a_1b_2c_3e_5| \end{vmatrix} . \qquad (8)$$

The derivation of further identities of the kind can be continued step by step. Clearing of fractions we obtain

$$\begin{aligned}
|a_1| \ |a_1b_2c_3| &= |\ |a_1b_2| \ |a_1c_3| \ |, \\
|a_1b_2| \ |a_1b_2c_3d_4| &= |\ |a_1b_2c_3| \ |a_1b_2d_4| \ |, \qquad (9) \\
|a_1b_2c_3| \ |a_1b_2c_3d_4e_5| &= |\ |a_1b_2c_3d_4| \ |a_1b_2c_3e_5| \ |,
\end{aligned}$$

and so on. Here the determinants of the 2nd order on the right have been indicated by their diagonal elements, which themselves are determinants. (Determinants having determinant elements are called compound determinants.) The identities are now polynomial identities, valid whether the determinants and minors concerned vanish or not. They are cases of more general identities (**45**) called extensional identities.

5. Expansion of a skew symmetric determinant of even order yields a polynomial which is a perfect square. For the identity (9) applied to the determinant gives on the left the product of the determinant and its leading minor, also skew symmetric, of order $n-2$, and on the right, as is easily seen (**19**, Ex. 9), a skew symmetric determinant of order 2. Hence,

since a skew symmetric determinant of order 2 is by inspection a square, the original determinant will be a square provided that its leading minor of order $n-2$ is a square. By the same reasoning this minor will be a square provided that the leading minor of order $n-4$ is a square. Proceeding in this way, we see that the result depends on the particular case, that a skew symmetric minor of order 2 is a square ; and the truth of this is evident.

In this proof we have assumed that the leading minors of even order do not vanish. No generality is really lost by this assumption.

The reader should go through the steps of the above proof, taking a skew symmetric determinant of the 4th or 6th order.

The square root of a skew symmetric determinant of even order is a polynomial in the $\frac{1}{2}n(n-1)$ elements, called a *Pfaffian*. Pfaffians have properties in some respects resembling those of determinants, but they lack a simple multiplication theorem, such as is derived for determinants in **34**.

6. The Pfaffian of order 2 is $a_{12}a_{34} - a_{13}a_{24} + a_{14}a_{23}$.

7. Prove that the terms in a Pfaffian are composed of factors a_{ij}, $i < j$, without repetition of suffixes. By enumeration of all such terms show that the number of terms in a Pfaffian of order n is $1 \cdot 3 \cdot 5 \ldots (2n-1)$.

Note on History of Determinants.—The standard works of reference, which the reader should consult for detailed information, are Muir's *History of the Theory of Determinants* (Macmillan), in four volumes, and *Contributions to the History of Determinants*, 1900-1920 (Blackie).

Vandermonde may be regarded as the founder (1771) of a notation and calculus of determinants. Of writers prior to 1850 Cauchy (1812), Schweins (1825) and Jacobi (1841) made specially important contributions in papers of these dates.

Matrices were introduced by Cayley in 1857.

CHAPTER III

THE ADJUGATE AND THE RECIPROCAL MATRIX: SOLUTION OF SIMULTANEOUS EQUATIONS: RANK AND LINEAR DEPENDENCE

21. The Adjugate Matrix of a Square Matrix

LET us substitute for the ith row of a determinant $|A|$ a new row of elements b_{ij}. The cofactors of the b_{ij} in the new determinant are evidently the same as those of the corresponding a_{ij} in $|A|$, and so an expansion of the new determinant is

$$b_{i1}|A_{i1}|+b_{i2}|A_{i2}|+\ldots+b_{in}|A_{in}|. \qquad . \qquad (1)$$

A similar result holds when a new j^{th} column is substituted in A.

Let us consider the case when the b_{ij} are elements of any row of A other than the ith. The expansion (1) is then the expansion of a determinant with two rows identical, and so gives zero. Thus we have the important result :

$$a_{k1}|A_{i1}|+a_{k2}|A_{i2}|+\ldots+a_{kn}|A_{in}| = 0, \quad k \neq i, \quad . \quad (2)$$

and a similar result is again true for expansions in terms of a column.

With respect to the elements of a given row (or column) of A we shall refer to the cofactors of elements in a different row (or column) as *alien* cofactors. Hence by (2) :

Expansions in terms of alien cofactors vanish identically.

The Adjugate Matrix. Let a matrix M be constructed, the elements of which are the cofactors $|A_{ij}|$

of elements a_{ij} of A, but placed in transposed position. Thus

$$M = [\,|A_{ij}|\,]' = [\,|A_{ji}|\,]. \qquad . \qquad . \qquad . \qquad (3)$$

This matrix is called the *adjugate* or *adjoint* of A, and will be noted by adj A. Its determinant $|\operatorname{adj} A|$ will be called the *adjugate determinant* of A, or the adjugate of $|A|$.

(In most books on determinants an untransposed adjugate is used, but for consistency and unification of theory it is preferable to transpose.)

1.
If $A = \begin{bmatrix} a_1 & b_1 & c_1 \\ a_2 & b_2 & c_2 \\ a_3 & b_3 & c_3 \end{bmatrix}$, adj $A = \begin{bmatrix} |b_2 c_3| & -|b_1 c_3| & |b_1 c_2| \\ -|a_2 c_3| & |a_1 c_3| & -|a_1 c_2| \\ |a_2 b_3| & -|a_1 b_3| & |a_1 b_2| \end{bmatrix}$

2.
If $A = \begin{bmatrix} 1 & 2 & 3 \\ 1 & 3 & 5 \\ 1 & 5 & 12 \end{bmatrix}$, adj $A = \begin{bmatrix} 11 & -9 & 1 \\ -7 & 9 & -2 \\ 2 & -3 & 1 \end{bmatrix}$.

3. Evaluate $|A|$ for the matrix of Ex. 2. Also form the product matrices $A(\operatorname{adj} A)$ and $(\operatorname{adj} A)A$. Try some other examples of the 3rd order.

Forming the product $A(\operatorname{adj} A)$, and keeping before us the case $n = 3$ for illustration,

$$\begin{bmatrix} a_{11} & a_{12} & a_{13} \\ a_{21} & a_{22} & a_{23} \\ a_{31} & a_{32} & a_{33} \end{bmatrix} \begin{bmatrix} |A_{11}| & |A_{21}| & |A_{31}| \\ |A_{12}| & |A_{22}| & |A_{32}| \\ |A_{13}| & |A_{23}| & |A_{33}| \end{bmatrix} = \begin{bmatrix} |A| & \cdot & \cdot \\ \cdot & |A| & \cdot \\ \cdot & \cdot & |A| \end{bmatrix}, (4)$$

we observe that by (2) all non-diagonal elements in the product vanish, being expansions in terms of alien cofactors, while the diagonal elements are expansions of $|A|$ in terms of elements of a row and proper cofactors. The reasoning is general for any order, and so we have

$$A(\operatorname{adj} A) = |A|\,I. \qquad . \qquad . \qquad . \qquad (5)$$

In the same way let us form $(\operatorname{adj} A)A$. The reader will assure himself that the diagonal elements are $|A|$, this time by expansion according to elements of a *column* and cofactors, and that all non-diagonal elements vanish as before. Hence

$$A \,(\text{adj } A) = (\text{adj } A)A = |A|I. \qquad . \qquad (6)$$

Singular and Nonsingular Matrices.—A square matrix A for which the determinant $|A| = 0$ is said to be *singular*. If $|A| \neq 0$, then A is said to be *nonsingular*.

By (6), if A is singular, $A(\text{adj } A) = (\text{adj } A)A = 0$. On the other hand, if A is nonsingular, we may divide the identity (6) through by $|A|$, obtaining $AR = RA = I$, where

$$R = |A|^{-1} \text{adj } A = [|A_{ji}|/ |A|] = [|A_{ij}|/ |A|]'. \qquad (7)$$

The matrix R thus disclosed may appropriately be named the *reciprocal matrix* of A, and may be denoted by A^{-1}. Its uniqueness is established in Ex. 6 below. It may be noted that it is both a pre-reciprocal and a post-reciprocal, a circumstance which depends, ultimately, on the fact that $|A'| = |A|$. We note also that it exists only when A is square and non-singular. Singular matrices have an adjugate, but no reciprocal.

At this stage the long-deferred operation of *division* for matrices may be introduced. If A is nonsingular we may operate with it in *predivision*, as in $A^{-1}B$, or in *post-division*, as in BA^{-1}, and just as in the case of multiplication these operations will in general produce different results.

4. If A is nonsingular and $AB = BA$, then $A^{-1}B = BA^{-1}$.

5. Evaluate $A^{-1}B$ and AB^{-1}, where

$$A = \begin{bmatrix} 1 & 2 & 3 \\ 1 & 3 & 5 \\ 1 & 5 & 12 \end{bmatrix}, B = \begin{bmatrix} 1 & 1 & 1 \\ 1 & 2 & 3 \\ 1 & 4 & 9 \end{bmatrix}.$$

6. Given only that the pre- and post-reciprocals of a matrix A exist, we may prove that they are identical and moreover unique. For suppose R and S are such that $RA = I = AS$. Then $R = RI = RAS = IS = S$. Further, if $TA = I$, then $R-T = (R-T)AS = S-S = 0$. Hence $R = T$. In the same way the post-reciprocal S is unique. These are simple consequences of $A = IA = AI$.

7. If $A, B, C, \ldots$ of order $n \times n$, are transformed to HAH^{-1}, HBH^{-1}, HCH^{-1}, $\ldots$ show that any polynomial in $A, B, C, \ldots$ with scalar coefficients is similarly transformed.

8. The reciprocal of a nonsingular diagonal matrix is a diagonal matrix. For example :

$$\begin{bmatrix} \lambda & . & . \\ . & \mu & . \\ . & . & \nu \end{bmatrix}^{-1} = \begin{bmatrix} \lambda^{-1} & . & . \\ . & \mu^{-1} & . \\ . & . & \nu^{-1} \end{bmatrix}, \quad \lambda\mu\nu \neq 0.$$

9. Let A be a matrix obtained by subjecting the rows of I to a certain permutation of order, and let B be the matrix obtained by subjecting the rows of I to the conjugate permutation. Prove that B and A are reciprocals. For example, corresponding to the conjugate permutations (2431) and (4132), the matrices below are reciprocal :

$$H = \begin{bmatrix} . & 1 & . & . \\ . & . & . & 1 \\ . & . & 1 & . \\ 1 & . & . & . \end{bmatrix}, \quad K = \begin{bmatrix} . & . & . & 1 \\ 1 & . & . & . \\ . & . & 1 & . \\ . & 1 & . & . \end{bmatrix}.$$

Note that $H' = K$, $H^{-1} = H'$.

10. The matrices $\begin{bmatrix} 1 & \lambda & \mu \\ . & 1 & . \\ . & . & 1 \end{bmatrix}$, $\begin{bmatrix} 1 & -\lambda & -\mu \\ . & 1 & . \\ . & . & 1 \end{bmatrix}$

are reciprocal. They each represent (19, Ex. 23) elementary operations.

Principal Minors. In a matrix A those minors which have elements situated symmetrically with respect to the principal diagonal of A are called principal minors. For example the diagonal elements a_{ii} are principal minors of order 1, their cofactors $|A_{ii}|$ in a determinant A are principal minors of order $n-1$, and $|A|$ itself may be regarded as the one principal minor of order n.

The principal minors of a Hermitian matrix are Hermitian. All principal minors in I are equal to 1, and all minors unsymmetrically placed are equal to 0.

22. Solution of Linear Equations in Nonsingular Case

The discovery of A^{-1} enables us to give the general solution of n linear equations in n unknowns when the matrix A of the equations is nonsingular. In fact, if the equations in matrix notation are

$$Ax = h, \text{ where } |A| \neq 0, \qquad . \qquad . \quad (1)$$

we have at once, premultiplying by A^{-1},

$$x = A^{-1}h, . \qquad . \qquad . \qquad . \quad (2)$$

which gives the vector of solutions. Further, since A^{-1} is unique, these are the only solutions.

To see how the set of solutions appears in ordinary notation and in terms of determinants and cofactors, let us keep again before us the case $n = 3$.

$$\begin{bmatrix} x_1 \\ x_2 \\ x_3 \end{bmatrix} = |A|^{-1} \begin{bmatrix} |A_{11}| & |A_{21}| & |A_{31}| \\ |A_{12}| & |A_{22}| & |A_{32}| \\ |A_{13}| & |A_{23}| & |A_{33}| \end{bmatrix} \begin{bmatrix} h_1 \\ h_2 \\ h_3 \end{bmatrix}. \quad (3)$$

Carrying out the multiplication on the right, and observing that every element of the vector obtained is by **21** (1) the expansion of a determinant according to elements h_i of a substituted column, we obtain

$$x_1 = \frac{\begin{vmatrix} h_1 & a_{12} & a_{13} \\ h_2 & a_{22} & a_{23} \\ h_3 & a_{32} & a_{33} \end{vmatrix}}{\begin{vmatrix} a_{11} & a_{12} & a_{13} \\ a_{21} & a_{22} & a_{23} \\ a_{31} & a_{32} & a_{33} \end{vmatrix}}, \quad x_2 = \frac{\begin{vmatrix} a_{11} & h_1 & a_{13} \\ a_{21} & h_2 & a_{23} \\ a_{31} & h_3 & a_{33} \end{vmatrix}}{\begin{vmatrix} a_{11} & a_{12} & a_{13} \\ a_{21} & a_{22} & a_{23} \\ a_{31} & a_{32} & a_{33} \end{vmatrix}}, \quad x_3 = \frac{\begin{vmatrix} a_{11} & a_{12} & h_1 \\ a_{21} & a_{22} & h_2 \\ a_{31} & a_{32} & h_3 \end{vmatrix}}{\begin{vmatrix} a_{11} & a_{12} & a_{13} \\ a_{21} & a_{22} & a_{23} \\ a_{31} & a_{32} & a_{33} \end{vmatrix}} \quad (4)$$

This result, which can clearly be extended to the general case, gives the rule of Cramer (1750), namely : each x_j is a quotient, the denominator of which is the determinant

$|A|$ of the system of equations, the numerator being a determinant obtained from $|A|$ by substituting the column vector $\{h_1 \ h_2 \ ... \ h_n\}$ for the jth column of $|A|$.

We may note at this point *three* equivalent methods of solving in practice a set of simultaneous equations. The first is the method of successive elimination followed by resubstitution ; the second is by evaluating the reciprocal matrix A^{-1} and operating with it upon the vector of constants on the right ; the third is by Cramer's rule, to evaluate the $n+1$ determinants of (4) and to form the n quotients. It will depend entirely on circumstances and on the means of computation available which method is used in a particular case.

1. Solve the equations

$$\begin{aligned} x+ \ y+ \ z &= \ 6, \\ x+2y+3z &= \ 14, \\ x+4y+9z &= \ 36 \end{aligned}$$

$x = 1$
$y = 2$
$z = 3$

by the various methods.

The reader should also construct various equations for himself, substitute numerical values of x, y, z to find the right-hand constants, and then solve, both by Cramer's rule and by forming the reciprocal matrix. The latter is a very simple procedure for a matrix of the 3rd order.

2. Premultiply by A' the set of equations $Ax = h$. The result is the set $A'Ax = A'h$, a set of equations with *symmetric* matrix $A'A$. Treat the set of equations of Ex. 1 in this way, and solve.

3. Practise finding the adjugate and reciprocal of various matrices of order 3×3 invented for the purpose. Notice that when the adjugate is found the determinant $|A|$ can be found in six different ways, by evaluating the diagonal elements of $A(\text{adj } A)$ and of $(\text{adj } A)A$. To use at least two such evaluations is to provide a useful check.

23. Reversal Rule for the Reciprocal of a Product Matrix

Reciprocal of Product Matrix. The reciprocal of a product matrix is the product of the reciprocals of the separate matrices in reversed order. For we have

$$AB \cdot B^{-1}A^{-1} = A \cdot I \cdot A^{-1} = AA^{-1} = I,$$
$$B^{-1}A^{-1} \cdot AB = B^{-1} \cdot I \cdot B = B^{-1}B = I.$$

Hence

$$(ABC)^{-1} = (AB \cdot C)^{-1} = C^{-1}(AB)^{-1} = C^{-1}B^{-1}A^{-1}, \quad (1)$$

and in the same way $(ABCD)^{-1} = D^{-1}C^{-1}B^{-1}A^{-1}$, and so on. All matrices concerned are of course nonsingular.

It is also easy to prove that the adjugate of a product matrix is the product of the separate adjugates in reversed order. Thus

$$\text{adj}\ (ABC) = \text{adj}\ C \cdot \text{adj}\ B \cdot \text{adj}\ A. \qquad (2)$$

This is only to be expected, since the adjugate has been defined in the first place as a *transposed* matrix, and so the reversal law for the transpose of a product (**9**) is operative.

Reciprocal Transformation. If $Ax = y$, $|A| \neq 0$, then $x = A^{-1}y$. Each of the linear transformations indicated by these equations is said to be the *reciprocal* of the other.

24. Orthogonal and Unitary Matrices

A square matrix A such that $A'A = AA' = I$, that is, $A' = A^{-1}$, is said to be *orthogonal*, for the following reason. Consider the distance from the origin to any point in n-space in rectangular Cartesian coordinates. The origin being denoted by the column vector $\{0\ 0\ \dots\ 0\}$ and the point by $x = \{x_1\ x_2\ \dots\ x_n\}$, the square of the distance is $x_1^2 + x_2^2 + \dots + x_n^2 = x'x$. This is unchanged by any rotation of axes about the fixed origin, provided that the axes remain rectangular, that is, orthogonal. Now the rotation is expressed by a linear transformation A, and its effect upon x is to produce the new coordinates Ax. The square of the distance from the origin is now $x'A'Ax$. But this is unchanged for *all* points x in the space. Hence we have $x'A'Ax = x'x$ identically in x, in other words the quadratic form $x'(A'A-I)x$ is identically zero. Hence $A'A-I = 0$, so that $A'A = I$.

Unitary Matrices. The extension of the concept of orthogonality to matrices with complex elements is effected by defining a *unitary* matrix A by the condition that $A^{-1} = \bar{A}'$, or $\bar{A}'A = A\bar{A}' = I$. A real unitary matrix is evidently an orthogonal matrix, so that all the properties of orthogonal matrices will be contained as special cases of properties of unitary matrices.

1. Prove that $A'A = AA' = I$, where

$$A = \begin{bmatrix} \cos\theta & \sin\theta \\ -\sin\theta & \cos\theta \end{bmatrix}.$$

This orthogonal matrix represents the change of rectangular axes in two dimensions by rotation through an angle θ.

2. Prove that the product of any number of unitary (and so of orthogonal) matrices is a unitary (or orthogonal) matrix.

For orthogonal matrices the geometrical interpretation is that a succession of rotations about a fixed origin has the same resultant effect as a single rotation.

3. Prove that the adjugate and the reciprocal of a Hermitian matrix are Hermitian.

4. Prove that the product of two Hermitian matrices is not in general Hermitian. Write down a few numerical symmetric matrices of the 3rd order, form their products in pairs and examine the results.

5. Prove that the matrix

$$\begin{bmatrix} 1/\sqrt{3} & (1+i)/\sqrt{3} \\ (1-i)/\sqrt{3} & -1/\sqrt{3} \end{bmatrix}$$

is unitary, where $i = \sqrt{-1}$.

6. Let L be the matrix of order 3×3 with rows

$$[l_1 \ m_1 \ n_1], \ [l_2 \ m_2 \ n_2], \ [l_3 \ m_3 \ n_3],$$

these being respectively the direction cosines of the new rectangular axes produced by a rotation in 3 dimensions of rectangular axes with fixed origin. Then L is orthogonal.

Write down in full the 12 relations given by the equalities $LL' = I$ and $L'L = I$, and observe that these are the "relations between direction cosines of three mutually perpendicular straight lines," as given in treatises on coordinate geometry of 3 dimensions.

In n dimensions there are $n(n+1)$ corresponding relations, all included in the matrix equality $LL' = L'L = I$.

7. Let the coordinates of n arbitrary points in n-dimensional Cartesian space be taken as the columns x_{ij} of a matrix X. Then if A is orthogonal we have $X'A'AX = X'X$, since $A'A = I$. The diagonal elements $x'_{ii}x_{ii}$ of $X'X$ are the squares of distances from the origin and are preserved invariant under the rotation A. But in virtue of $X'A'AX = X'X$ the *non-diagonal* elements $x'_{ii}x_{ij}$ are also preserved invariant. Their geometrical interpretation is the product of the distances of the points x_{ii} and x_{ij} from the origin, multiplied by the *cosine* of the angle between the lines joining those points to the origin; in other words, the product of either line and its projection on the other. In two dimensions, for example, we have

$$x_{i1} = \{r_1\cos\theta \ \ r_1\sin\theta\}, \ x_{i2} = \{r_2\cos\phi \ \ r_2\sin\phi\}$$

and the relation in question is

$$[r_1\cos\theta \ \ r_1\sin\theta]\{r_2\cos\phi \ \ r_2\sin\phi\} = r_1 r_2 \cos (\theta - \phi).$$

25. The Solution of Homogeneous Linear Equations

In the equations $Ax = h$ all the elements of Ax are of the first degree in the elements of x, but the constants on the right, the elements of h, are of zero degree in the elements of x. For this reason the equations $Ax = h$ are described as *non-homogeneous*. On the other hand, the equations $Ax = 0$ are *homogeneous*. We now inquire what conditions must be satisfied by A in order that the equations $Ax = 0$ may have a non-trivial solution $x \neq 0$; for we reject as trivial the obvious solution $x = 0$. It is clear that the vector x, if it exists, will be arbitrary to the extent at least of a scalar constant factor, for if $Ax = 0$ then certainly $Ay = 0$, where $y = \lambda x$. For this investigation, as well as for the later discussion of m non-homogeneous equations in n unknowns, we require the important concepts of *rank* and *linear dependence*.

26. Rank and Nullity of a Matrix

Rank. If A is a matrix, rectangular or square, and if all minors of order $r+1$ contained in it are zero while at least one minor of order r is not zero, then A is said to be of *rank r*.

Nullity. If A is square of order $n \times n$, then $n-r$, the complement of the rank with respect to the order, is called the *nullity* of A. If A is rectangular of order $m \times n$ there are two nullities, a *row-nullity $m-r$* and a *column-nullity $n-r$*.

1. A row or column vector with at least one nonzero element is of rank 1.

2. The unit matrix of order $n \times n$ has rank n and nullity 0.

3. The null matrix of order $m \times n$ has rank 0, row-nullity m, column-nullity n.

4. The matrix of order $m \times n$ in which every element is unity has rank 1, row-nullity $m-1$, column-nullity $n-1$.

5. If u and x are nonzero row and column vectors of orders $1 \times n$ and $n \times 1$, show that the scalar ux and the matrix xu are of rank 1 or 0. Examine some numerical examples.

6. The matrix U of order $n \times n$ has units (**11**, Ex. 5) in the first superdiagonal and zeros elsewhere. Show that the rank of U^k is $n-k$, $k \leqslant n$.

Elementary operations on rows or columns of A, involving permutations of order or linear combinations, the modified row or column receiving a unit or at least nonzero coefficient in any such operation, leave the rank of A invariant. For under any permutation of rows or columns the minors of A will receive at most a change of sign, but not of numerical value ; and under an operation

$$\text{row}_i + \lambda_1\text{row}_1 + \lambda_2\text{row}_2 + \ldots + \lambda_k\text{row}_k . \qquad . \quad (1)$$

any minor $|A_i|$ containing elements of row_i will become

$$|A_i| \pm \lambda_1|A_1| \pm \lambda_2|A_2| \pm \ldots \pm \lambda_k|A_k|, \qquad . \quad (2)$$

where $|A_1|$, $|A_2|$, ..., $|A_k|$ are minors obtained from $|A_i|$ by substituting in row_i the corresponding elements of row_1, row_2, ..., row_k respectively. This follows at once from (1) by expanding the modified minor in terms of row_i. Similar remarks apply to operations of linear combination on columns. It follows that if all minors of order $r+1$ in A are zero, then all minors of order $r+1$ in PAQ, where P and Q are the nonsingular matrices (**19**, Ex. 19, 20, 21, 22, 23) effecting these elementary operations, are also zero. In other words, if $B = PAQ$ then B cannot have higher rank than A. But P^{-1} and Q^{-1} are also (**21**, Ex. 9) elementary operations of the same kind ; and so since $A = P^{-1}BQ^{-1}$, A cannot have higher rank than B.

Hence A and B must have the same rank r.

27. Linear Dependence of Functions, Vectors and Matrices

Linear Dependence. If a set of functions $f_1, f_2, \ldots, f_n$ or of vectors $f_{(1}, f_{(2}, \ldots, f_{(n}$ or of matrices $F_1, F_2, \ldots, F_n$ satisfies identically a linear relation

$$c_1f_1 + c_2f_2 + \ldots + c_nf_n = 0 \quad . \qquad . \quad (1)$$

or the like with $f_{(j}$ or F_j for f_j, where the scalar constants c_j are *not all zero*, then the functions or vectors or matrices are said to be *linearly dependent*. More precisely any f affected by a nonzero coefficient c_j in (1) is said to be linearly dependent on the remaining f's. If no relation such as (1) is possible the f_j are said to be *linearly independent*.

1. The linear dependence of the functions in (1) can be expressed concisely in the form $c'f = 0$, $c' \neq 0$, where c' is the row vector of scalars c_j, f the column vector of functions.

2. The linear dependence of vectors may be expressed in matrix notation. For example, if $a_{(1}$, $a_{(2}$, $a_{(3}$ are column vectors, linearly dependent in virtue of a relation

$$x_1 a_{(1} + x_2 a_{(2} + x_3 a_{(3} = 0 \qquad \cdot \qquad \cdot \qquad \cdot \qquad (2)$$

this relation is exactly the same (24, Ex. 1) as

$$Ax = [a_{(1} a_{(2} \, a_{(3}] \{x_1 \, x_2 \, x_3\} = 0, \qquad \cdot \qquad (3)$$

where A is a partitioned matrix of which the vectors $a_{(j}$ are columns, and $x \neq 0$. In the same way, to assert that certain row vectors $a'_{(i}$ are linearly dependent is to assert that the homogeneous equations $uA = 0$, where the row vectors $a'_{(i}$ are the rows of A, have a non-trivial solution $u \neq 0$.

3. If a matrix A is partitioned by rows into $\{A_1 \, A_2\}$, and every row of the submatrix A_2 is linearly dependent on the rows of A_1, then $A_2 = CA_1$, where the elements in the successive rows of C are the respective coefficients in the successive relations of dependence. (The reader should write out in full some examples of low order.)

4. If the rows of a square matrix A are linearly dependent then $|A| = 0$; for there exists an operation on rows, namely, that given by the relation of dependence, which will replace all the elements of some row by zeros without altering the value of $|A|$. Naturally the coefficient of this particular row in the relation of dependence must be unity.

For example, if the rows of a matrix A of order 3×3 satisfy the relation $\text{row}_1 - 2 \, \text{row}_2 - 3 \, \text{row}_3 = 0$, then the operation $\text{row}_1 - 2 \, \text{row}_2 - 3 \, \text{row}_3$ applied to $|A|$ reduces all elements of the first row to zero; hence $|A| = 0$.

5. For the same reason if the columns of A are linearly dependent then $|A| = 0$.

Let us insist that we are not yet entitled to assert the converse theorems, namely, that if $|A| = 0$ the rows and the columns of A are linearly dependent.

6. If every set of r rows of a rectangular matrix A is linearly

dependent then A is of rank $r-1$ or less, since by Ex. 4 all minors of order r in A must vanish. Similarly for columns.

28. Conditions for Solution of Homogeneous Equations

Let the equations be in n unknowns x_i, and m in number. In fact we consider $Ax = 0$, where A is of order $m \times n$ and of rank r.

Necessary Condition. The necessary condition for $Ax = 0$, $x \neq 0$ is that r shall be less than n, the number of unknowns.

For suppose $r = n$, which can happen only if $m \geqslant n$. Then by a rearrangement, if necessary, of the equations, that is, of the rows of A, we can make the submatrix A_1 constituted by the first n rows nonsingular. Partitioning it from the submatrix B_1 constituted by the remaining $m-n$ rows, we have the equations in partitioned form

$$\begin{bmatrix} A_1 \\ B_1 \end{bmatrix} x = \begin{bmatrix} 0 \\ 0 \end{bmatrix}, \text{ so that } A_1 x = 0, B_1 x = 0. \quad . \quad (1)$$

Now since A_1 is nonsingular A_1^{-1} exists ; and so $A_1 x = 0$ leads to

$$A_1^{-1} A_1 x = x = 0,$$

contrary to the hypothesis that $x \neq 0$. Hence the condition $r < n$ is necessary.

Sufficient Condition. The same condition is also sufficient. For suppose $r < n$, and let the equations and also the unknowns be arranged, as is always possible, so that the leading submatrix A_1 of order $r \times r$ is nonsingular. The leading submatrix of order $(r+1) \times (r+1)$ will of course be singular. Let the cofactors of elements of the last row, the $(r+1)^{th}$, of this singular submatrix be taken as the first $r+1$ elements of a column vector x, and let the last $n-r-1$ elements of x be zeros. Then $x \neq 0$, since its $(r+1)^{th}$ element is $|A_1| \neq 0$. We shall proceed to prove that x, so constructed, satisfies $Ax = 0$.

It may be helpful to the reader to have an illustration under view, in which $m = 4$, $n = 5$, $r = 2$.

$$Ax \equiv \begin{bmatrix} a_1 & b_1 & c_1 & d_1 & e_1 \\ a_2 & b_2 & c_2 & d_2 & e_2 \\ a_3 & b_3 & c_3 & d_3 & e_3 \\ \hdashline a_4 & b_4 & c_4 & d_4 & e_4 \end{bmatrix} \begin{bmatrix} |b_1 c_2| \\ -|a_1 c_2| \\ |a_1 b_2| \\ \cdot \\ \cdot \end{bmatrix} \equiv \begin{bmatrix} \cdot \\ \cdot \\ \cdot \\ \cdot \end{bmatrix} = 0. \quad (2)$$

All elements in Ax are zero, either because they are expansions by alien cofactors (**21**, (2)) or because they are expansions of minors of order $r+1$ taken from the first r rows and some later row of A ; and all such minors vanish. (The reader should verify this for every element of the product in the illustration.) Hence a solution $x \neq 0$ has been constructed ; and so the condition $r < n$ is sufficient.

1. If A is of order $n \times n$ and $|A| = 0$, then r must be $< n$. Hence $Ax = 0$, $x \neq 0$ is possible. In other words (**27**, Ex. 2) the columns of A are linearly dependent. In the same way, since $|A'| = |A| = 0$, the rows of A are linearly dependent.

The determinant $|A|$ is often called the eliminant of the equations $Ax = 0$.

Thus we have the important converse of **27**, Ex. 4, 5, namely : the vanishing of a determinant both implies, and is implied by, the linear dependence of its rows and of its columns.

This converse theorem enables us to discuss rank by means of linear dependence. For if B is a rectangular matrix of rank r, it contains a non-vanishing minor $|A|$ of order r. Hence the r rows (or columns) of B which contain $|A|$ are linearly independent. On the other hand, any $r+1$ rows (or columns) of B are linearly dependent, since otherwise they would contain some non-vanishing minor of order $r+1$, contrary to hypothesis. The important conclusion is that the statements (i) B is of rank r, (ii) the maximum number of linearly independent rows or columns in B is r, are equivalent.

2. A symmetric matrix of rank r contains at least one non-vanishing *principal* minor of order r. For since the matrix is of rank r, a certain r rows must be linearly independent. Hence elementary operations exist which annul the remaining rows and, by symmetry, the corresponding columns. Hence the minor contained by the r rows and columns is nonzero ; and it is evidently a principal minor.

The extension to the case of a Hermitian matrix is evident.

3. If $m > n$, any set of m vectors of n elements is linearly dependent. For suppose them to be column vectors of a matrix, and annex $m - n$ rows of zero elements to make a larger matrix A of order $m \times m$. Then $|A| = 0$. Hence the columns of A are linearly dependent ; and so the original column vectors (namely, the columns of A, each stripped of its last $m - n$ zeros) are linearly dependent.

4. If $x_1, x_2, \ldots, x_n$ are scalar numbers *all different*, the n vectors $[1 \ x_j \ x_j^2 \ldots x_j^{n-1}]$ are linearly independent.

5. Let A be of order $n \times (n+1)$ and of rank n, the leading submatrix A_1 of order $n \times n$ being nonsingular. Then the $(n+1)^{th}$ column of A is either itself null, or is linearly dependent on the preceding n columns, as follows :

Let A be written in partitioned form $[A_1 \ a]$, where a is the last column, and consider

$$[A_1 \ a] \begin{bmatrix} x \\ -1 \end{bmatrix} = 0, \text{ that is, } A_1 x = a.$$

Since $|A_1| \neq 0$ the equations $A_1 x = a$ have a unique solution $x = A_1^{-1} a$. The relation $A_1 x = a$ is then (**27** (3)) the desired relation of dependence of columns. The reader should try some numerical examples, such as

$$A = \begin{bmatrix} 1 & 2 & 1 & -1 \\ 1 & 3 & 3 & 2 \\ 1 & 4 & 6 & -3 \end{bmatrix}.$$

6. It follows from Ex. 5 that any nonzero column vector of n elements can be expressed as a linear combination of the columns of a nonsingular matrix A of order $n \times n$; with a corresponding result for any nonzero row vector.

E

7. The vectors of the nonsingular matrix are said to constitute a *basis*, or *vector basis*, in terms of which an arbitrary vector of the same order can be linearly expressed. The simplest basis is provided by the rows

$$e_{\{1} = [1 \; 0 \; 0 \; ... \; 0], \; e_{\{2} = [0 \; 1 \; 0 \; ... \; 0], \; ..., \; e_{\{n} = [0 \; 0 \; ... \; 0 \; 1]$$

of the unit matrix I, and in the same way the columns of I. For example

$$u = [u_1 \; u_2 \; u_3] = u_1[1 \; 0 \; 0] + u_2[0 \; 1 \; 0] + u_3[0 \; 0 \; 1]. \quad . \quad (3)$$

The solution which we constructed at (2) for $Ax = 0$, $r < n$ is in general not unique, since *any* column later than the rth could have been moved into the $(r+1)^{th}$ position, thus producing in general a different leading submatrix of order $(r+1) \times (r+1)$ and therefore different cofactors as the elements of x; for example, elements d_i or e_i instead of c_i in the solving vector of (2). The solution is therefore arbitrary to an extent which now calls for consideration.

29. Reduction of a Matrix to Equivalent Form

Let A be a matrix of order $m \times n$ and of rank r. By elementary operations, rearrangements of rows and columns if necessary, we may bring a nonsingular submatrix A_1 of order $r \times r$ into leading position. The operations being equivalent to nonsingular pre- and postmultiplications (**19**, Ex. 19 to 23) which do not alter rank (**26**), we have

$$PAQ = \begin{bmatrix} A_1 & B_1 \\ A_2 & B_2 \end{bmatrix}, \quad . \quad . \quad . \quad (1)$$

where P and Q are nonsingular matrices of order $m \times m$ and $n \times n$, and PAQ is of rank r. Since A_1 has a reciprocal A_1^{-1} the following further reductions are possible, as the reader should verify by carrying out the partitioned multiplications indicated :

$$\begin{bmatrix} I & \cdot \\ -A_2 A_1^{-1} & I \end{bmatrix} \begin{bmatrix} A_1 & B_1 \\ A_2 & B_2 \end{bmatrix} = \begin{bmatrix} A_1 & B_1 \\ \cdot & B_2 - A_2 A_1^{-1} B_1 \end{bmatrix}, \quad (2)$$

$$\begin{bmatrix} A_1 & B_1 \\ A_2 & B_2 \end{bmatrix} \begin{bmatrix} I & -A_1^{-1} B_1 \\ \cdot & I \end{bmatrix} = \begin{bmatrix} A_1 & \cdot \\ A_2 & B_2 - A_2 A_1^{-1} B_1 \end{bmatrix}, \quad (3)$$

$$\begin{bmatrix} I & \cdot \\ -A_2 A_1^{-1} & I \end{bmatrix} \begin{bmatrix} A_1 & B_1 \\ A_2 & B_2 \end{bmatrix} \begin{bmatrix} I & -A_1^{-1} B_1 \\ \cdot & I \end{bmatrix}$$

$$= \begin{bmatrix} A_1 & \cdot \\ \cdot & B_2 - A_2 A_1^{-1} B_1 \end{bmatrix}. \quad \cdot \quad \cdot \quad \cdot \quad (4)$$

Now $B_2 - A_2 A_1^{-1} B_1$ must be null, because if it contained any nonzero element c_{ij} then the submatrix, on the right of (2) or (3) or (4), consisting of A_1 bordered by the row and column containing c_{ij} would be nonsingular, having a determinant $c_{ij} |A_1| \neq 0$. But then the rank would be $r+1$ or more, infringing the theorem that the rank of A is invariant under the elementary operations (**19**, Ex. 23) which have been used.

Hence we have the important results, that if A is of rank r it can be reduced by equivalent nonsingular transformations PA, AQ, PAQ to respective forms

$$\begin{bmatrix} A_1 & B_1 \\ \cdot & \cdot \end{bmatrix}, \quad \begin{bmatrix} A_1 & \cdot \\ A_2 & \cdot \end{bmatrix}, \quad \begin{bmatrix} A_1 & \cdot \\ \cdot & \cdot \end{bmatrix}, \quad |A_1| \neq 0, \quad \cdot \quad (5)$$

where $|A_1|$ is of order r.

We may call the first two of these *semi-reduced* forms. By removing a further nonsingular factor

$$\begin{bmatrix} A_1 & \cdot \\ \cdot & I \end{bmatrix} \quad \cdot \quad \cdot \quad \cdot \quad \cdot \quad (6)$$

and absorbing it into either P or Q we have the still simpler equivalent reductions

$$\begin{bmatrix} I & A_1^{-1} B_1 \\ \cdot & \cdot \end{bmatrix}, \quad \begin{bmatrix} I & \cdot \\ A_2 A_1^{-1} & \cdot \end{bmatrix}, \quad \begin{bmatrix} I & \cdot \\ \cdot & \cdot \end{bmatrix}, \quad \cdot \quad (7)$$

where the leading submatrix I is of order $r \times r$.

It is also to be noted that the elements of the matrices P and Q which effect these reductions PAQ are all *rational* functions of the elements of A. This follows from inspection of (2), (3), (4) and (6).

1. If A is symmetric it contains (**28**, Ex. 2) a non-vanishing principal minor. By a rearrangement of rows, and exactly the same rearrangement of columns, this minor may be brought into leading position. The transformation effecting this is of congruent type $P'AP$. The transformation of **29** (4) may then be applied ; but we may observe that because of the symmetry of A we must have $A'_1 = A_1$, $B_1 = A'_2$, and so $(A_1^{-1}B_1)' = A_2A_1^{-1}$. It follows that in this case the transformation (4) is again congruent, and so for symmetric matrices we derive the fully reduced form in (5) by a congruent transformation of type $P'AP$.

2. The reader should go through the same steps for a Hermitian matrix A, deriving the fully reduced form (5) by a conjunctive transformation of type $\overline{P}'AP$.

The first semi-reduction in (5) enables us to discuss the extent of the arbitrariness in the solution of a system of homogeneous equations. We may suppose the equations and the unknowns rearranged so that the leading submatrix A_1 of the system is of order $r \times r$ and nonsingular. Applying the semi-reduction, we may write the equations $Ax = 0$ in the partitioned shape

$$P^{-1}\begin{bmatrix} A_1 & B_1 \\ . & . \end{bmatrix}\begin{bmatrix} x_{(1} \\ x_{(2} \end{bmatrix} = \begin{bmatrix} 0 \\ 0 \end{bmatrix}, \qquad \cdot \qquad \cdot \qquad (8)$$

where x has been partitioned into its first r and last $n-r$ elements. Hence

$$\begin{bmatrix} A_1 & B_1 \\ . & . \end{bmatrix}\begin{bmatrix} x_{(1} \\ x_{(2} \end{bmatrix} = \begin{bmatrix} 0 \\ 0 \end{bmatrix}, \text{ or } A_1x_{(1} = -B_1x_{(2}, \qquad (9)$$

so that $\qquad\qquad x_{(1} = -A_1^{-1}B_1x_{(2}. \qquad \cdot \qquad \cdot \qquad \cdot \qquad (10)$

This form of the solution shows that the $n-r$ elements of $x_{(2}$ can be assigned quite arbitrarily, and that when this is done the r elements of $x_{(1}$ are uniquely determined. Hence the important result :

If A is of rank r less than n, the number of unknowns, and if the columns of A corresponding to a certain r unknowns constitute a matrix of rank r, then the remaining $n-r$ unknowns can be assigned arbitrarily, and may be regarded as parameters in terms of which the other r unknowns can be linearly and uniquely expressed.

3. If
$$x_1 + 2x_2 + 3x_3 + 4x_4 = 0$$
$$x_1 + 3x_2 + 5x_3 + 7x_4 = 0,$$

express x_1 and x_2 in terms of x_3 and x_4 ; also express x_1 and x_3 in terms of x_2 and x_4.

We further observe (**28, Ex. 6**) that the number of linearly independent vectors $x_{(2}$ cannot exceed $n-r$. Hence from the form of the solutions of $Ax = 0$, namely

$$x = \begin{bmatrix} -A_1^{-1}B_1 x_{(2} \\ x_{(2} \end{bmatrix}, \qquad . \qquad . \qquad (11)$$

it is equally clear that the number of linearly independent solutions x also cannot exceed $n-r$.

30. Consistency and Solution of Non-Homogeneous Equations

The non-homogeneous equations $Ax = h$ can be treated in a similar manner by the semi-reduction of **29** (5), which gives

$$\begin{bmatrix} A_1 & B_1 \\ . & . \end{bmatrix} \begin{bmatrix} x_{(1} \\ x_{(2} \end{bmatrix} = P \begin{bmatrix} h_{(1} \\ h_{(2} \end{bmatrix} = \begin{bmatrix} k_{(1} \\ k_{(2} \end{bmatrix}, \qquad . \qquad . \qquad (1)$$

let us say. Carrying out the multiplication on the left we observe that $k_{(2}$ must be zero. If $k_{(2} \neq 0$ the equations $Ax = h$ cannot be solved ; they are then said to be *incompatible* or *inconsistent*.

The condition for consistency, $k_{(2} = 0$, can be expressed in a more direct way. If we compare the partitioned matrices

$$\begin{bmatrix} A_1 & B_1 \\ \cdot & \cdot \end{bmatrix} \text{ and } \begin{bmatrix} A_1 & B_1 & k_{(1} \\ \cdot & \cdot & k_{(2} \end{bmatrix}, \qquad . \qquad . \quad (2)$$

we may note that if $k_{(2} = 0$ they are of the same rank r, whereas if $k_{(2} \neq 0$ the second matrix is of rank $r+1$, since at least one submatrix of order $(r+1) \times (r+1)$, containing A_1 and some nonzero element k_i of $k_{(2}$, is nonsingular, its determinant being $k_i |A_1| \neq 0$. Hence for compatibility of the equations $Ax = h$ the two matrices in (2) must have the same rank r. Premultiplying by P^{-1}, which does not alter the rank, we have the condition that A and $[A \ h]$ must have the same rank r. It is customary to call $[A \ h]$, constructed as it is by adjoining to A the column of constants on the right of the equations, the *augmented matrix* of the system. Hence the criterion:

The necessary condition for the consistency of a set of non-homogeneous linear equations is that both the matrix of the system and the augmented matrix should have the same rank.

1. Investigate the consistency of the equations

$$\begin{aligned} x_1 + \ x_2 &= 4, \\ x_1 + 2x_2 &= 7, \\ 2x_1 + 5x_2 &= 15 \end{aligned}$$

Inconsistent
Rank of $A = 2$
$Ah = 3$

by the above test of rank, and also from first principles.

If the condition of consistency is satisfied we have

$$\begin{bmatrix} A_1 & B_1 \\ \cdot & \cdot \end{bmatrix} \begin{bmatrix} x_{(1} \\ x_{(2} \end{bmatrix} = \begin{bmatrix} k_{(1} \\ \cdot \end{bmatrix}, \qquad . \qquad . \quad (3)$$

so that $\qquad A_1 x_{(1} + B_1 x_{(2} = k_{(1}, \qquad . \qquad . \quad (4)$

yielding $\qquad x_{(1} = -A_1^{-1} B_1 x_{(2} + A_1^{-1} k_{(1}. \qquad . \quad (5)$

Here again, just as in the case of homogeneous equations, the values of the $n-r$ unknowns in $x_{(2}$ can be assigned arbitrarily, and when this is done the r elements of $x_{(1}$ are uniquely determined. Hence we have another important result :

If a set of non-homogeneous equations in n unknowns is consistent, the rank of the matrix and of the augmented matrix being r, then the values of $n-r$ of the unknowns may be assigned arbitrarily, provided that the set of columns corresponding to the remaining r unknowns forms a submatrix of rank r.

2. Test for consistency and solve :

$$x_1 + x_2 + x_3 = 6,$$
$$x_1 + 2x_2 + 3x_3 = 14,$$
$$x_1 + 4x_2 + 7x_3 = 30.$$

Criteria for Rank. The following examples embody results which serve as criteria for the rank of a matrix.

3. If all the rows (or columns) of a matrix A are linearly dependent on a certain r rows (or columns) which are themselves linearly independent, then A is of rank r.

Partition A into $\{A_1 \ A_2\}$, where A_1 contains the linearly independent r rows, A_2 the remainder. Then **(27, Ex. 3)** we must have $A_2 = CA_1$. Accordingly

$$\begin{bmatrix} I & \vdots \\ -C & I \end{bmatrix} \begin{bmatrix} A_1 \\ A_2 \end{bmatrix} = \begin{bmatrix} A_1 \\ \cdot \end{bmatrix},$$

that is, an elementary nonsingular operation has reduced A to a matrix which is evidently of rank r. Hence **(26)** A is of rank r.

4. If a certain minor $|A_1|$ of order r in A is nonzero, while all minors of order $r+1$ containing $|A_1|$ are zero, then A is of rank r.

Without loss of generality, we may take $|A_1|$ as leading minor. Then **(28, Ex. 1)** the first r rows of A are linearly independent. Let any other row of A be adjoined as $(r+1)^{th}$ row. We shall prove that it is linearly dependent on the first r rows.

For since by hypothesis the leading minor of order $r+1$ is zero, its $(r+1)^{th}$ row must be linearly dependent on its first r rows by a relation

$$\text{row}_{r+1} + c_1\text{row}_1 + c_2\text{row}_2 + \ldots + c_r\text{row}_r = 0. \quad . \quad (6)$$

If we substitute for the $(r+1)^{th}$ column of this leading minor the corresponding elements of some later column, we obtain another minor of order $r+1$ which contains $|A_1|$ and so vanishes. Evaluating this minor by applying the operation on the left of (6) we reduce the first r elements of its $(r+1)^{th}$ row to zero ; but the $(r+1)^{th}$ element must also be zero, because if it were $c \neq 0$ then the value of the minor would be $c|A_1| \neq 0$, contrary to hypothesis. (*Cf.* **29** (4).)

It follows that the operation on the left of (6) reduces *all* elements of the $(r+1)^{th}$ row of A to zero ; in other words, the $(r+1)^{th}$ row of A is linearly dependent on the first r rows, (6) being the relation of dependence.

Hence every row of A is linearly dependent on the first r rows, which are themselves linearly independent. Hence, by Ex. 3, A is of rank r.

The practical consequence is that rank may be ascertained by pivotal condensation, for by the condensation process (**20**) applied to A each pivot is contained in its successors, and the pivots involve minors of A of successively higher orders. It follows from this and from Ex. 4 that if we regard the setting up of A as stage 1, the first condensation with respect to a pivot in A as stage 2, the second condensation as stage 3, and so on, then the last stage possessing non-vanishing elements is stage r. This is perhaps the most expeditious way of ascertaining the rank of a matrix.

5. If A is symmetric and possesses a non-vanishing principal minor $|A_1|$ of order r, such that all principal minors of orders $r+1$ and $r+2$ containing $|A_1|$ vanish, then A is of rank r.

Without disturbing symmetry we may take $|A_1|$ as leading minor, and we may bring any two principal minors of orders $r+1$ and $r+2$ containing $|A_1|$ into leading position also.

Let us exemplify the three minors by $|a_1b_2c_3|$, $|a_1b_2c_3d_4|$, $|a_1b_2c_3d_4e_5|$. By **20**, (8)

$$|a_1b_2c_3d_4e_5|\ |a_1b_2c_3| = \begin{vmatrix} |a_1b_2c_3d_4| & |a_1b_2c_3e_4| \\ |a_1b_2c_3d_5| & |a_1b_2c_3e_5| \end{vmatrix},$$

that is,
$$0 = -\ |a_1b_2c_3e_4|^2,$$

since $|a_1b_2c_3d_4e_5| = 0$, $|a_1b_2c_3d_4| = 0$ and by symmetry $|a_1b_2c_3d_5| = |a_1b_2c_3e_4|$. Hence $|a_1b_2c_3e_4|$, a non-diagonal minor of order $r+1$ containing $|A_1|$, vanishes ; and so it follows that all such non-diagonal minors, as well as the principal ones, vanish. Hence, by Ex. 4, A is of rank r.

The reader will see that the same theorem holds for Hermitian matrices.

* Latent Vectors, Latent Roots. The vectors and scalars introduced below are fundamental in matrix theory.

6. If A is of order $n \times n$, row vectors u and column vectors q exist such that $uA = \lambda u$, $Aq = \lambda q$. For $u(A - \lambda I) = 0$ and $(A - \lambda I)q = 0$, if $|A - \lambda I| = 0$. This is the *characteristic equation* of A, of degree n in λ. Its n roots λ_j are the *latent roots* of A ; and so to each λ_j corresponds a pair of *latent vectors* $u_{(j}$, $q_{(j}$, determined but for a scalar factor.

7. Prove that $|B - \lambda I| = |A - \lambda I|$, where $B = KAK^{-1}$. Thus B and A have the same characteristic equation and latent roots. The matrix K is nonsingular but arbitrary.

8. If A is Hermitian $Aq = \lambda q$ gives $\bar{q}'A = \bar{\lambda}\bar{q}'$. Thus each $\bar{q}'$ is a latent row vector u.

9. The latent roots of a Hermitian matrix are real. For $Aq = \lambda q$ gives $\bar{q}'Aq = \lambda\bar{q}'q$. The Hermitian forms (**10**, Ex. 13) $\bar{q}'Aq$, $\bar{q}'q$ are real, and $\bar{q}'q$ is not zero. Hence λ is real.

10. The latent roots of a unitary matrix have unit modulus. For $Aq = \lambda q$ gives $\bar{q}'\bar{A}'Aq = \bar{\lambda}\lambda\bar{q}'q$. But $\bar{q}'\bar{A}'Aq = \bar{q}'q$, since $\bar{A}'A = I$. Cancelling the nonzero $\bar{q}'q$ we obtain $\bar{\lambda}\lambda = 1$.

The reader should find the latent roots in **24**, Ex. 1, 5.

11. Establish step by step $A^2q = \lambda^2 q$, ..., $A^kq = \lambda^k q$. Also $A^{-1}q = \lambda^{-1}q$ for nonsingular A. Deduce that if $\psi(\lambda)$ is a rational function, $\psi(A)$ nonsingular, then $\psi(A)q = \psi(\lambda)q$.

* In a first reading this section may be omitted.

CHAPTER IV

FURTHER EXPANSIONS : CAUCHY AND LAPLACE EXPANSIONS : MULTIPLICATION THEOREMS

31. Expansion of Determinant by Elements of a Row and a Column

THE expansion of a determinant $|A|$ according to elements of a row and their cofactors, or elements of a column and their cofactors, is only one of many possible expansions.

Consider for example the evident identity

$$\begin{vmatrix} a_{11} & a_{12} & a_{13} \\ a_{21} & a_{22} & a_{23} \\ a_{31} & a_{32} & a_{33} \end{vmatrix} = a_{11}|A_{11}| + \begin{vmatrix} 0 & a_{12} & a_{13} \\ a_{21} & a_{22} & a_{23} \\ a_{31} & a_{32} & a_{33} \end{vmatrix}, \qquad (1)$$

in which the second determinant on the right is what is called a *bordered* determinant, if we regard it as being formed by bordering $|a_{22} a_{33}|$ with a prefixed row $[0\ a_{12}\ a_{13}]$ and column $\{0\ a_{21}\ a_{31}\}$. It is evident that such an identity holds for a determinant $|A|$ of any order.

Now let us expand the second term on the right of (1), the bordered determinant, according to elements of its first row a_{12}, a_{13}, ..., a_{1n}. Every cofactor $|A_{12}|$, $|A_{13}|$, ..., $|A_{1n}|$ contains the elements a_{21}, a_{31}, ..., a_{n1} of the first column. Let us then expand all these cofactors in their turn according to these elements a_{21}, a_{31}, ..., a_{n1}, denoting the cofactor of a_{i1} in $|A_{1j}|$ by $|A_{1j;\,i1}|$, that is, $-|A_{11;\,ij}|$. The aggregate of terms so obtained is

$$|A| = a_{11}|A_{11}| - \underset{i}{\sum}\underset{j}{\sum} a_{i1}a_{1j}|A_{11;\,ij}|, \quad i, j \neq 1. \qquad (2)$$

This is an expansion of $|A|$, due to Cauchy, in terms of elements of a row *and* of a column and the joint cofactors, which apart from the first term will be signed minors of order $n-2$. Clearly the expansion need not be restricted to elements of the first row and first column; a corresponding expansion in terms of the hth row and k^{th} column could have been established, namely,

$$|A| = a_{hk}|A_{hk}| - \underset{i\ j}{\Sigma\Sigma}a_{ik}a_{hj}|A_{hk;\ ij}|, \quad i \neq h, j \neq k. \quad (3)$$

The reader may easily construct another proof of this important expansion by an enumeration of terms thus: first show that all the terms in the expansion are different and are also terms of $|A|$, then show that the number of terms arising is $(n-1)! + (n-1)^2(n-2)!$, which reduces to $n!$.

1. Let a matrix A of order $n \times n$ be bordered by a first row $\{0\ x_1\ x_2 \ldots x_n\}$ and a first column $[0\ y_1\ y_2 \ldots y_n]$. Then the expansion of its determinant in terms of the elements of the first row and column will be

$$- \underset{i\ j}{\Sigma\Sigma}x_iy_j|A_{ji}|, . \qquad . \qquad . \qquad (4)$$

a bilinear form which in matrix notation would be written as $-x'(\text{adj } A)y$. Thus if $x'(\text{adj } A)y$ is called the *adjoint* or *adjugate* bilinear form of $x'Ay$, we have it in the form of a bordered determinant

$$- \begin{vmatrix} . & x' \\ y & A \end{vmatrix}. \qquad . \qquad . \qquad . \qquad (5)$$

2. If A is symmetric and $y = x$ the above bilinear form becomes the adjoint or adjugate quadratic form of $x'Ax$, namely, $x'(\text{adj } A)x$. If A is Hermitian and $y = \bar{x}$, we obtain the adjoint Hermitian form of $\bar{x}'Ax$.

The reader should try some numerical examples of low order.

3. If $A = I$, the quadratic form and the adjoint quadratic form are the same, namely, the sum of squares $x'x$.

4. If A is symmetric and nonsingular, express the reciprocal quadratic form $x'A^{-1}x$ as a bordered determinant.

5. Evaluate several determinants of the 4th order with numerical elements by Cauchy's expansion in terms of elements of a row and of a column. Observe carefully the signs of the terms, and check by alternative evaluations.

32. Complementary Minors : Algebraic Complements or Cofactors

Let us take the general determinant $|A|$ of order n and partition off its first m rows and first m columns. The remaining $n-m$ rows and columns constitute a submatrix of order $(n-m)\times(n-m)$ which would provide a minor of order $n-m$, namely,

$$|a_{m+1,\,m+1}\ a_{m+2,\,m+2} \cdots a_{nn}|, \qquad . \qquad . \quad (1)$$

as represented by its diagonal elements. Now the coefficient of $a_{11}a_{22}\ldots a_{mm}$ in $|A|$ is evidently

$$\Sigma \pm a_{m+1,\sigma}a_{m+2,\tau} \cdots a_{n\nu}, \qquad . \qquad . \quad (2)$$

where, since the row suffixes $1, 2, \ldots, m$ and column suffixes $1, 2, \ldots, m$ are in natural order, the sign of any term depends on the permutation $(\sigma\,\tau\ldots\nu)$ of the natural order $(m+1,\ m+2,\ \ldots,\ n)$. But the aggregate of such terms is merely the minor (1), which is said to be the *complementary minor* of $|a_{11}a_{22}\ldots a_{mm}|$. In general two minors of orders m and $n-m$ formed from complementary sets of rows and columns in A are said to be complementary with respect to A.

Next, while leaving unaltered the last $n-m$ columns of A, we may systematically permute the first m columns in such a way as to bring $a_{1\alpha}a_{2\beta} \cdots a_{m\mu}$ into the diagonal, where $(\alpha\,\beta\ldots\mu)$ is any assigned permutation of $(123\ldots m)$. Such operations merely affect $|A|$ with the sign proper to this permutation. It follows that the complementary minor $|a_{m+1,\ m+1}a_{m+2,\ m+2}\ldots a_{nn}|$ is not merely the coefficient of $a_{11}a_{22}\ldots a_{mm}$ in $|A|$; it is the coefficient of

$$\Sigma \pm a_{1\alpha}a_{2\beta} \cdots a_{m\mu}, \qquad . \qquad . \quad (3)$$

that is, of $|a_{11}a_{22}\ldots a_{mm}|$, its own complementary minor. In the same way, as we can see at once by bringing *any* minor of order m into leading position, the product of any two complementary minors, affected with the sign produced by bringing either of them into leading position, yields terms occurring in $|A|$.

To discover this sign, consider the minor taken from rows i_1, i_2, ..., i_m and columns j_1, j_2, ..., j_m of A, both of these sets being in the usual ascending order. By

$$(i_1-1)+(i_2-2)+ \ldots +(i_m-m) \qquad . \qquad . \quad (4)$$

interchanges of rows, and

$$(j_1-1)+(j_2-2)+ \ldots +(j_m-m) \qquad . \qquad . \quad (5)$$

interchanges of columns such a minor can be brought into leading position. Hence in the expansion of $|A|$ the corresponding product of complementary minors must take on the sign factor due to

$$i_1+j_1+i_2+j_2+ \ldots +i_m+j_m-m(m+1)$$

sign changes, and since $m(m+1)$ is even it can be discarded. Hence the desired sign factor is

$$(-)^{i_1+j_1+i_2+j_2+ \ldots +i_m+j_m} \qquad . \qquad . \qquad . \quad (6)$$

An alternative way of finding the sign is by inspection of diagonal suffixes. For example $|a_{12}a_{24}a_{45}|$ and its complement $|a_{31}a_{53}|$ show row suffixes (12435), column suffixes (24513). The relative class (**16**) is even, and so the desired sign is positive.

The complementary minor of a given minor, with the above sign prefixed, is often called the *algebraic complement* of the given minor in $|A|$. There is no reason why it should not be called its *cofactor* in $|A|$, and we shall adopt this name. Any minor, with its sign quâ cofactor prefixed, may also be called a *signed minor*, as in **19**.

Leaving a minor in its original position in $|A|$, obliterate the remaining elements of $|A|$. The rows and columns thus rendered blank constitute a submatrix. Insert unit elements successively in the vacated rows so that this submatrix now becomes a unit submatrix. Prove, by bringing the submatrix of units into leading position, that the determinant so obtained is equal to the signed minor corresponding to the original minor. For example

$$\begin{vmatrix} \cdot & 1 & \cdot & \cdot & \cdot \\ a_{21} & \cdot & a_{23} & \cdot & \cdot \\ \cdot & \cdot & \cdot & 1 & \cdot \\ \cdot & \cdot & \cdot & \cdot & 1 \\ a_{51} & \cdot & a_{53} & \cdot & \cdot \end{vmatrix} = - \begin{vmatrix} a_{21} & a_{23} \\ a_{51} & a_{53} \end{vmatrix}.$$

33. The Laplacian Expansion of a Determinant

Let us take any m rows of $|A|$; no generality will be lost by taking the first m rows. From these we may form $n_{(m)}$ minors of order m, where $n_{(m)} = n(n-1)\ldots(n-m+1)/m!$. Multiplying each minor by its cofactor, a signed minor of order $n-m$, we obtain terms of $|A|$. None of these terms duplicates any other, for they involve different selections of rows and columns, and so each term has a different arrangement of suffixes from every other term. Counting up all the terms of $|A|$, so arising, we have altogether

$$n_{(m)} m!(n-m)! = \frac{n!}{m!(n-m)!} m!(n-m)! = n! \quad . \quad (1)$$

terms. Hence we have *all* the terms of $|A|$ without omission or repetition.

This type of expansion of $|A|$, according to minors from a certain m rows and their cofactors from the remaining $n-m$ rows, was first given by Laplace in 1772. Since $|A'| = |A|$, it is evident that expansion by minors and cofactors from complementary sets of columns is equally possible.

1. $|a_1 b_2 c_3 d_4| = |a_1 b_2| \ |c_3 d_4| - |a_1 c_2| \ |b_3 d_4| + |a_1 d_2| \ |b_3 c_4|$
$\qquad + |b_1 c_2| \ |a_3 d_4| - |b_1 d_2| \ |a_3 c_4| + |c_1 d_2| \ |a_3 b_4|,$

by a Laplacian expansion according to the first two and the last two rows of the determinant.

The reader is invited to write down other expansions with various choices of two rows or columns and their complementaries.

2. Construct various determinants of order 4 with numerical elements and evaluate them by Laplacian expansion, checking the results by Cauchy's expansion.

A quasi-Laplacian expansion, according to minors from a certain set of rows (the same applies to columns) and the cofactors of corresponding minors not from that set of m rows but from another set of m rows, vanishes identically. This is the expansion by *alien* cofactors. The reason is that in any such expansion the elements of at least one row of $|A|$ must occur twice, once in the minors of order m and once in the cofactors of other minors by which these are multiplied. Therefore the expansion is that of a determinant derived from $|A|$ in which at least two rows are identical ; and such a determinant vanishes.

This result is the natural extension of our earlier one, namely that an expansion according to elements of a row of $|A|$ and alien cofactors (21) vanishes identically.

3. In the determinant $|a_1 b_2 c_3 d_4 e_5 f_6|$ put $[a_4\ b_4\ c_4\ d_4]$, $[a_5\ b_5\ c_5\ d_5]$, $[a_6\ b_6\ c_6\ d_6]$ equal to $[a_1\ b_1\ c_1\ d_1]$, $[a_2\ b_2\ c_2\ d_2]$, $[a_3\ b_3\ c_3\ d_3]$ respectively ; also put $\{e_4\ e_5\ e_6\}$ and $\{f_4\ f_5\ f_6\}$ equal to zero. The determinant then vanishes. Expanding it in terms of its first three and last three rows, we derive the identity

$$|a_1b_2c_3||d_1e_2f_3| - |a_1b_2d_3||c_1e_2f_3| + |a_1c_2d_3||b_1e_2f_3| - |b_1c_2d_3||a_1e_2f_3|$$
$$= 0.$$

4. By modifying a determinant $|a_1b_2c_3d_4e_5f_6g_7h_8|$ in a similar way, so that the last four rows repeat the first four except for the fact that $\{g_5\,g_6\,g_7\,g_8\}$ and $\{h_5\,h_6\,h_7\,h_8\}$ are null, and by expanding the resulting determinant in terms of its first four and last four rows, we obtain an identity not unlike the above, but of 15 terms, each being a product of two determinants of the 4th order. The reader should obtain this identity.

This method of Laplacian expansion applied to vanishing determinants may be made to yield many such identities in sums of products of determinants. The identities are of importance in the theory of algebraic invariants.

Before we pass to more general expansions of Laplace and Cauchy type, we may prove by the ordinary Laplacian expansion the fundamental theorem of multiplication of two determinants.

34. Multiplication of Determinants

The multiplication theorem for determinants can be expressed in relation to matrix theory as follows :

The determinant of the product of two square matrices is equal to the product of the determinants of those matrices ; and similarly for the product of any number of square matrices. In symbols, $|AB...K| = |A| \, |B| ... |K|$.

Proof. Consider two square matrices A and B of order $n \times n$, and take them as diagonal submatrices of a partitioned matrix of order $2n \times 2n$; of the remaining non-diagonal submatrices let the one above the diagonal be null and the one below the diagonal be $-I$. The following identity is then readily verified, all matrices being conformably partitioned :

$$\begin{bmatrix} I & A \\ \cdot & I \end{bmatrix} \begin{bmatrix} A & \cdot \\ -I & B \end{bmatrix} = \begin{bmatrix} \cdot & AB \\ -I & B \end{bmatrix} \qquad . \qquad . \quad (1)$$

(The reader is invited to write this identity in full non-partitioned form, taking for A and B square matrices of the 3rd order, and to refer to this illustration at the various stages of the proof.)

Now the operation of premultiplication on the left of (1) is of a kind already encountered (**19**, Ex. 23) and is equivalent to row operations which do not change the value of a determinant. Hence, taking determinants, we have

$$\begin{vmatrix} A & \cdot \\ -I & B \end{vmatrix} = \begin{vmatrix} \cdot & AB \\ -I & B \end{vmatrix} , \qquad . \qquad . \quad (2)$$

and expanding each side by a Laplacian expansion according to the first n rows, we have the desired result

$$|A| \, |B| = |AB|, \qquad . \qquad . \qquad . \quad (3)$$

the positive sign of $|AB|$ being assured by an inspection of leading terms in $|A|$, $|B|$, and $|AB|$. We deduce at once

$$|ABC| = |A| \ |BC| = |A| \ |B| \ |C|, \qquad . \qquad (4)$$

and so on for any number of matrices of order $n \times n$.

Not only so, but since $|A'| = |A|$ we have the equally possible equivalents $|AB'|$, $|A'B|$, $|A'B'|$ for $|A| \ |B|$; in other words, determinant multiplication can be carried out row into row, column into column, and column into row, as well as by the customary row into column way of matrices.

1. The reader should here take various determinants of the 3rd and 4th order which he has already evaluated for practice, should multiply their matrices in the four ways just mentioned and should verify that the determinants of the results are indeed the products of the separate determinants.

2. Since $A(\text{adj } A) = |A|I$ we obtain, taking determinants, $|A| \ |\text{adj } A| = |A|^n$. Hence if A is nonsingular $|\text{adj } A| = |A|^{n-1}$, a theorem due to Cauchy. The theorem expresses a polynomial relation which persists even when it is singular. In that case $|A| = 0$, $|\text{adj } A| = 0$.

3. Prove that sums and products of matrices of order 2×2 of the type

$$A = \begin{bmatrix} a_{11} & a_{12} \\ -\bar{a}_{12} & \bar{a}_{11} \end{bmatrix}$$

are matrices of the same type. By taking the determinant of a product AB of such matrices, prove that the product of two sums of four squares can be expressed rationally in certain ways as the sum of four squares, and find those ways.

35. Extensions of the Laplace and Cauchy Expansions

In the terms of a Laplacian expansion the minors of order m, or their cofactors of order $n - m$, can themselves be further expanded in Laplacian fashion, as a result of which we have an aggregate of terms, each of which is a product of three non-overlapping minors; and these in their turn

F

can be further expanded. We arrive in this way at the following generalization of the Laplacian expansion :

Suppose that the n rows of A are partitioned into a certain n_1 rows, then another n_2 rows, ... and finally the n_k rows left over, where

$$n_1 + n_2 + \ldots + n_k = n.$$

The rows in any set need not be consecutive rows in A. Let minors, of respective orders n_1, n_2, ... n_k, be taken in all possible ways from the respective sets of rows ; and let products be formed containing one minor from each set in such a way that no column of A enters twice, in other words, all columns are used, sign being given by consideration of the diagonal elements of minors in each product. Then the sum of all possible such signed products of k minors is an expansion of $|A|$.

If this were not otherwise evident from the description of the way in which this expansion is derived by repeated Laplacian expansion, it could be made evident by enumeration of terms ; for the number of terms of $|A|$ arising from expanding all these products, which themselves number

$$n_{(n_1)} \, (n-n_1)_{(n_2)} \, (n-n_1-n_2)_{(n_3)} \ldots (n-n_1- \ldots -n_{k-1})_{(n_k)}, \qquad (1)$$

will be

$$\frac{n!}{n_1!(n-n_1)!} \, \frac{(n-n_1)!}{n_2!(n-n_1-n_2)!} \ldots \frac{(n-n_1-\ldots-n_k)!}{n_k!} n_1! n_2! \ldots n_k! = n!, \quad (2)$$

and this is the number of terms in $|A|$.

In particular, if $k = n$ and $n_1 = n_2 = \ldots = n_k = 1$, the expansion is merely the sum of $n!$ signed products of single elements, the defining expansion of $|A|$.

1. Construct some determinants of the 5th order with numerical elements. Partition the rows or columns in various ways according to the partition 2, 2, 1 of 5 and evaluate the determinants by the extended Laplacian expansion, checking the results by alternative evaluations.

2. Every expansion of $|A|$ that we have so far used is equally applicable to the permanent of A, the sole difference being that all terms and all permanent minors receive positive sign. The reason is that the argument used to demonstrate the expansions is an enumerative one, which applies equally well to permanents. For example

$$\overset{+}{\begin{vmatrix} a_1 & b_1 & c_1 \\ a_2 & b_2 & c_2 \\ a_3 & b_3 & c_3 \end{vmatrix}}\overset{+}{} = a_1 \overset{+}{\begin{vmatrix} b_2 c_3 \end{vmatrix}} \overset{+}{+} b_1 \overset{+}{\begin{vmatrix} a_2 c_3 \end{vmatrix}} \overset{+}{+} c_1 \overset{+}{\begin{vmatrix} a_2 b_3 \end{vmatrix}} \overset{+}{}. \quad (3)$$

and similar expansions.

Extended Cauchy Expansion. Cauchy's expansion of $|A|$ in terms of the elements of a row and of a column, which is in essence the expansion of a bordered determinant, can be extended to the case where m rows and m columns are used in bordering. The bordered determinant is then of the form

$$\begin{vmatrix} \cdot & X \\ Y' & A \end{vmatrix}, \text{ where } X = \begin{bmatrix} x_{11} & x_{12} & x_{13} & x_{14} \\ x_{21} & x_{22} & x_{23} & x_{24} \\ x_{31} & x_{32} & x_{33} & x_{34} \end{bmatrix}, \quad (4)$$

for example, and Y resembles X. In the general case A is of order $n \times n$, X of order $m \times n$ and so Y' of order $n \times m$.

It is evident by Laplacian expansion according to the first m rows of (4) that if $m = n$ the value of the bordered determinant is $(-)^n |X| \, |Y|$, and that if $m > n$ the determinant vanishes.

When $m < n$ a Laplacian expansion of $n_{(m)}$ terms is possible, by minors of order m in X each multiplied by its cofactor of order n in the bordered determinant. Further, every cofactor contains all the elements of Y'. Let us expand in turn all of these cofactors according to the first m columns containing Y'. When this is done the previous Laplacian expansion yields an expansion in which every term has *three* factors, the first being a minor of order m taken from X, the second a minor of order m taken from Y', and the third a minor of order $n - m$ taken from A and involving those rows and columns of A not included in the other two minors. The sum of all possible products, $(n_{(m)})^2$ in number, which can be formed in this way, correct sign being given by reference to diagonal elements in the factors, constitutes an expansion of the bordered determinant which generalizes the ordinary Cauchy expansion according to one row and one column.

A further extension of similar nature concerns a bordered determinant resembling that in (4), but with A of order $n \times p$, X of order $m \times p$, $m < p$, Y of order $n \times (m+n-p)$.

1. Show that the number of terms in the final expansion in products of elements of the bordered determinant is $(n!)^2/(n-m)!$, and show that the number of terms given by the extended Cauchy expansion is the same as this.

2. The extended Cauchy expansion applies equally well to a permanent, all products receiving positive sign.

3. Expand by the extended Cauchy expansion

$$\begin{vmatrix} \cdot & X \\ Y' & A \end{vmatrix} = \begin{vmatrix} \cdot & \cdot & x_1 & x_2 & x_3 \\ \cdot & \cdot & u_1 & u_2 & u_3 \\ y_1 & v_1 & a_1 & b_1 & c_1 \\ y_2 & v_2 & a_2 & b_2 & c_2 \\ y_3 & v_3 & a_3 & b_3 & c_3 \end{vmatrix},$$

paying careful attention to the sign of the nine terms.

4. In Ex. 3 put $A = -I$ and observe the form of the result. Also put $A = -I$ and $Y = X$ and again observe the form of the result. Generalize these results.

5. In Ex. 3 give numerical values to the elements in various ways, and evaluate the bordered determinants thus constructed.

6. Write out completely the expansion the first term of which is given below, paying careful attention to the sign of terms.

$$\begin{vmatrix} \cdot & x_1 & x_2 & x_3 & x_4 \\ \cdot & u_1 & u_2 & u_3 & u_4 \\ v_1 & a_1 & a_2 & a_3 & a_4 \\ v_2 & b_1 & b_2 & b_3 & b_4 \\ v_3 & c_1 & c_2 & c_3 & c_4 \end{vmatrix} = |x_1 u_2| |v_1| |b_3 c_4| + \cdots .$$

36. Determinant of a Product of
Rectangular Matrices

Consider AB, where A is of order $m \times n$ and B is of order $n \times m$. The determinant $|AB|$ is of order m. We can evaluate it by making partitioned matrices exactly as in our proof of the multiplication theorem of determinants (34 (1)), and using the same identity

$$\begin{bmatrix} I & A \\ \cdot & I \end{bmatrix} \begin{bmatrix} A & \cdot \\ -I & B \end{bmatrix} = \begin{bmatrix} \cdot & AB \\ -I & B \end{bmatrix}. \qquad (1)$$

We have therefore

$$\begin{vmatrix} A & \cdot \\ -I & B \end{vmatrix} = \begin{vmatrix} \cdot & AB \\ -I & B \end{vmatrix}. \qquad (2)$$

exactly as before, but since A and B are no longer square but rectangular the expansions of the determinants and the signs of terms need more attention.

The Laplacian expansion of the determinant on the right is $|AB|$, with sign factor. The determinant on the left can be expanded by the extended Cauchy expansion, A and B being regarded as bordering $-I$. It may help to keep under view the particular examples given below.

$$\text{(i)} \begin{vmatrix} a_{11} & a_{12} & \cdot & \cdot & \cdot \\ a_{21} & a_{22} & \cdot & \cdot & \cdot \\ a_{31} & a_{32} & \cdot & \cdot & \cdot \\ -1 & \cdot & b_{11} & b_{12} & b_{13} \\ \cdot & -1 & b_{21} & b_{22} & b_{23} \end{vmatrix}, \quad \text{(ii)} \begin{vmatrix} a_{11} & a_{12} & a_{13} & \cdot & \cdot \\ a_{21} & a_{22} & a_{23} & \cdot & \cdot \\ -1 & \cdot & \cdot & b_{11} & b_{12} \\ \cdot & -1 & \cdot & b_{21} & b_{22} \\ \cdot & \cdot & -1 & b_{31} & b_{32} \end{vmatrix}. \quad (3)$$

In the case $m > n$ the expansion is identically zero, as we have seen. Hence if $m > n$ we have $|AB| = 0$.

1. This could equally well be proved by adding $m - n$ zero columns to A to make a *square* matrix $[A\ 0]$, likewise $m - n$ zero rows to B, and observing that

$$[A\ 0] \begin{bmatrix} B \\ 0 \end{bmatrix} = AB.$$

The ordinary multiplication theorem of determinants now gives $|AB| = 0$.

On the other hand, if $m < n$, since the only nonzero minors of any order in I, or $-I$, are its diagonal or principal minors of that order, the Cauchy expansion (**35**, Ex. 4) in question consists of these minors of order $n-m$ multiplied by their cofactors in A and B together. From the *symmetry* of $-I$, if the cofactor in A comes from a certain m columns, the cofactor in B will come from the *corresponding* m rows. Thus in (ii), fixing sign by leading terms, we have

$$|AB| = |a_{11}a_{22}|\ |b_{11}b_{22}| + |a_{11}a_{23}|\ |b_{11}b_{32}| + |a_{12}a_{23}|\ |b_{21}b_{32}|. \quad (4)$$

The general result may be stated thus :

If A is of order $m \times n$ and B of order $n \times m$, $m \leqslant n$, then $|AB|$ is equal to the sum of $n_{(m)}$ products made by pairing each minor of order m from m columns of A with the minor of order m from the *corresponding rows* of B. All such products have the same sign, since AB is unchanged when columns of A and rows of B are *similarly* permuted. If $m > n$, $|AB| = 0$. If $m = n$, $|AB| = |A|\ |B|$, as in **34** (3).

This fundamental theorem concerning the determinant of a product of rectangular matrices was discovered independently by Binet and Cauchy (1812) and is usually called the theorem of *multiplication of determinantal arrays*. It can be extended step by step to the case of the determinant of the product of any number of matrices, provided that the product is square. The enunciation of this very general theorem is as follows :

Let the determinant formed of elements common to rows α_1, α_2, ..., α_n and columns β_1, β_2, ..., β_n of a matrix T be denoted by

$$T\begin{pmatrix} \alpha_1\ \alpha_2\ ...\ \alpha_n \\ \beta_1\ \beta_2\ ...\ \beta_n \end{pmatrix}. \qquad \qquad (5)$$

Then if $T = AB...RS$, where A, B, ..., R, S are of order $k \times m$, $m \times n$, ..., $r \times s$, $s \times k$ respectively, so that T is of order $k \times k$, then

$$T = \underset{\beta\gamma}{\Sigma\Sigma}\cdots\underset{\sigma}{\Sigma}\ A\begin{pmatrix} 1\,2\ ...k \\ \beta_1\beta_2...\beta_k \end{pmatrix} B\begin{pmatrix} \beta_1\beta_2...\beta_k \\ \gamma_1\gamma_2...\gamma_k \end{pmatrix}...R\begin{pmatrix} \rho_1\rho_2...\rho_k \\ \sigma_1\sigma_2...\sigma_k \end{pmatrix} S\begin{pmatrix} \sigma_1\sigma_2...\sigma_k \\ 1\ 2\,...k \end{pmatrix}, (6)$$

where the summations are over all sets of k columns taken independently from the *columns* of $A, B, ..., R$; or, alternatively, over all sets of k rows taken independently from the *rows* of $B, ..., R, S$. The number of terms in the expansion is therefore

$$m_{(k)}n_{(k)}...r_{(k)}s_{(k)}, \qquad . \qquad . \qquad . \qquad (7)$$

and the expansion will vanish identically if any of the following inequalities be true,

$$k>m, \ k>n, \ ..., \ k>r, \ k>s.$$

2. Show that
$$\begin{vmatrix} \cdot & X' \\ X & A \end{vmatrix},$$

where X is of order $m \times n$, $m>n$, and A is symmetric, is a certain quadratic form, with the minors of X of order n as variables. If $m = 4$, $n = 2$, write down the matrix of the quadratic form. How is it formed from A?

37. Expansion of Determinant by Diagonal Elements

Let us consider the determinant of $A+X$, where X is a diagonal matrix $X = [x_{ii}]$; for example, to take the case $n = 4$,

$$A+X = \begin{bmatrix} a_{11}+x_{11} & a_{12} & a_{13} & a_{14} \\ a_{21} & a_{22}+x_{22} & a_{23} & a_{24} \\ a_{31} & a_{32} & a_{33}+x_{33} & a_{34} \\ a_{41} & a_{42} & a_{43} & a_{44}+x_{44} \end{bmatrix}. \qquad (1)$$

The determinant $|A+X|$ is evidently a polynomial in $x_{11}, x_{22}, x_{33}, x_{44}$. Expanding in terms of products of these variables and their coefficients or cofactors, we have

$$\begin{aligned} |A+X| = \ &x_{11}x_{22}x_{33}x_{44} + a_{11}x_{22}x_{33}x_{44} + a_{22}x_{11}x_{33}x_{44} + ... \\ &+ |a_{11}a_{22}|x_{33}x_{44} + |a_{11}a_{33}|x_{22}x_{44} + ... \\ &+ |a_{11}a_{22}a_{33}|x_4 + |a_{11}a_{22}a_{44}|x_3 + ... \\ &+ |a_{11}a_{22}a_{33}a_{44}|. \end{aligned} \qquad (2)$$

In general the expansion consists of a sum of products of x_{ii} taken m at a time, each product to be multiplied by the complementary principal minor of order $n-m$ in $|A|$; the sum being taken over such products for all values of m from n to 0. Permanents admit a similar expansion.

1. Expand the determinant $|A|$ of the 4th order in this diagonal fashion by regarding its matrix as

$$\begin{bmatrix} \cdot & a_{12} & a_{13} & a_{14} \\ a_{21} & \cdot & a_{23} & a_{23} \\ a_{31} & a_{32} & \cdot & a_{34} \\ a_{41} & a_{42} & a_{43} & \cdot \end{bmatrix} + \begin{bmatrix} a_{11} & \cdot & \cdot & \cdot \\ \cdot & a_{22} & \cdot & \cdot \\ \cdot & \cdot & a_{33} & \cdot \\ \cdot & \cdot & \cdot & a_{44} \end{bmatrix}$$

and expanding in terms of products of the a_{ii}.

2. Try this mode of expansion on some determinants of the 3rd and 4th order with numerical elements, checking the result by evaluation in other ways.

Spur or Trace of Matrix. A specially interesting and important case arises when X is not merely diagonal but scalar, $X = \lambda I$. We then derive from (2) the expansion

$$|A+\lambda I| = \lambda^n + \mathrm{sp}_1 A.\, \lambda^{n-1} + \mathrm{sp}_2 A.\, \lambda^{n-2} + \ldots + \mathrm{sp}_{n-1} A.\, \lambda + |A|, \quad (3)$$

where $\mathrm{sp}_r A$ means " sum of all principal minors of $|A|$ of order r ". The first of these sums, $\mathrm{sp}_1 A$, the sum of the diagonal elements a_{ii} themselves, is important in matrix theory and is often called the *spur* or *trace* of A. We shall denote it by $\mathrm{sp}\, A$. (Some authors employ $\mathrm{tr}\, A$.)

3. Expand

$$\begin{vmatrix} \lambda & -a & -b \\ a & \lambda & -c \\ b & c & \lambda \end{vmatrix}$$

in powers of λ.

* 4. Let $\phi(\lambda) = |A - \lambda I| = (-)^n(\lambda - \lambda_1)(\lambda - \lambda_2) \ldots (\lambda - \lambda_n)$. We may also expand $\phi(\lambda)$ in powers of λ, in the manner of (3). Equating coefficients of powers of λ, we deduce that $\mathrm{sp}_k A$ is the sum of products (*cf.* 48) of λ_j taken k at a time. In particular $\mathrm{sp}\, A$ is the sum, and $|A|$ the product, of the latent roots. Hence if any λ_j is zero A is singular ; and conversely.

* Normal or Canonical Form of a Matrix. In the next few examples we assume the λ_j to be all distinct. The case of repeated roots is important, but beyond our scope.

5. The latent vectors u or q of A are linearly independent. For if there were a relation with nonzero c_j, such as
$$p = c_1 q_{(1} + c_2 q_{(2} + \ldots + c_k q_{(k} = 0,$$
then taking p, Ap, $A^2 p$, ..., $A^{k-1}p$ we have by **30**, Ex. 11,
$$c_1 \lambda_1^s q_{(1} + c_2 \lambda_2^s q_{(2} + \ldots + c_k \lambda_k^s q_{(k} = 0, \quad s = 0, 1, \ldots, k-1.$$
These are sets of homogeneous equations in corresponding elements of the $q_{(j}$. For them to have nonzero solutions it is necessary (**28**) that $|\lambda_1^0 \lambda_2^1 \ldots \lambda_k^{k-1}| = 0$. But (**19**, Ex. 11) such a difference-product of distinct λ_j cannot vanish. The assumed relation is thus impossible.

6. The $u_{(i}$ and $q_{(j}$ are thus rows and columns of nonsingular matrices U and Q, and so the respective sets of relations $u_{(i}A = \lambda_i u_{(i}$, $Aq_{(j} = \lambda_j q_{(j}$ can be written as $UA = \Lambda U$, $AQ = Q\Lambda$, where Λ is the diagonal matrix $[\lambda_i]$.

7. Hence $UAU^{-1} = \Lambda = Q^{-1}AQ$, so that A has been reduced by the *similar* transformation (of type HAH^{-1}) to diagonal *canonical* form Λ. (For multiple λ_j this *may* not be possible.)

8. We may prove that $u_{(i}q_{(i} \neq 0$, $u_{(i}q_{(j} = 0$, $i \neq j$. For $u_{(i}Aq_{(j} = \lambda_i u_{(i}q_{(j} = \lambda_j u_{(i}q_{(j} = 0$, since $\lambda_i \neq \lambda_j$. Hence the non-diagonal elements of UQ are zero. Since UQ is nonsingular the diagonal elements $u_{(i}q_{(i} \neq 0$. Fixing the arbitrary scalars in $u_{(i}$ and $q_{(i}$ so that $u_{(i}q_{(i} = 1$, we have $UQ = I$.

9. If A is Hermitian each $u = \bar{q}'$, by **30**, Ex. 8. Hence $U = \bar{Q}'$, and so $UQ = I$ yields $\bar{Q}'Q = I$, whence Q is unitary. It follows that $\bar{Q}'AQ = \Lambda$, a unitary reduction to real diagonal canonical form; real, since the latent roots (**30**, Ex. 9) are real. In particular if A is real it is symmetric, and Q is orthogonal.

The quadratic form $x'Ax$ thus becomes $y'\Lambda y$, a form involving squares only, if $y = Q^{-1}x = Q'x$. (Orthogonal reduction of a central quadric to principal axes.)

10. The relations $\psi(A)q_{(j} = \psi(\lambda_j)q_{(j}$ of **30**, Ex. 11 can be assembled, in view of **7**, Ex. 10, into a single matrix relation $\psi(A)Q = Q\psi(\Lambda)$. Putting $\psi(\lambda) = \phi(\lambda) = |A - \lambda I|$, we have $\phi(A)Q = 0$. Since Q is nonsingular, $\phi(A) = 0$. This is the Cayley-Hamilton theorem, valid also for multiple roots.

* In a first reading these examples may be omitted.

CHAPTER V

COMPOUND MATRICES AND DETERMINANTS : DUAL THEOREMS

38. Compounds and Adjugate Compounds of a Matrix

THE elements of adj A are the cofactors of elements a_{ji} in $|A|$; and the special properties of adj A are intimately related to the expansion of $|A|$ according to elements of a row or of a column and their cofactors in $|A|$. The more general Laplacian expansion, in terms of minors of order k taken from a certain k rows or columns and their cofactors of order $n-k$, suggests the following extension of the adjugate relationship.

Compound Matrices. Let a matrix be formed the elements of which are minors of $|A|$ of order k ; let all minors which come from the same group of k rows (or columns) of A be placed in the same row (or column) of this derived matrix ; and let the priority of elements in rows or columns of this matrix be decided on the principle by which words are ordered in a dictionary or lexicon. For example, minors from rows 1, 2, 4, of A will appear in an earlier row than those from 1, 2, 5 or 1, 3, 4 or 2, 3, 4 ; and similarly for columns. We shall call this order *lexical* order. (The usual name is "lexicographical".) The matrix with elements minors of order k constructed in this way will be called the kth *compound* of A and will be denoted by $A^{(k)}$. It will be defined in the same way even when A is rectangular of order $m \times n$, submatrices of order $k \times k$ being extracted from A and their determinants being taken as the elements of $A^{(k)}$, lexical ordering being observed. The order of $A^{(k)}$ will then be $m_{(k)} \times n_{(k)}$.

Adjugate Compound Matrices. When A is square of order $n \times n$, every minor of order k in $|A|$ is accompanied, in a Laplacian expansion of $|A|$, by its cofactor or signed minor of order $n-k$. Let us take the kth compound $A^{(k)}$, replace every element in it by its cofactor in $|A|$, and transpose the resulting matrix. We thus obtain a matrix of the same order as $A^{(k)}$, which may be called the kth *adjugate compound* of A, and will be denoted by $\mathrm{adj}^{(k)}A$. We may notice in passing that $A^{(1)} = A$, $\mathrm{adj}^{(1)}A = \mathrm{adj}\,A$.

1. Take the general matrix of order 5×5 and write out *in extenso*, indicating minors by diagonal elements, the 2nd compound and 2nd adjugate compound, and the 3rd compound and 3rd adjugate compound, comparing the 2nd adjugate compound with the 3rd compound, and the 3rd adjugate compound with the 2nd compound. The comparison will reveal three features of difference.

2. The kth compound of I is a unit matrix. The kth compound of a diagonal matrix D is a diagonal matrix, with the k-ary products of the diagonal elements d_{ii}, in lexical order, as diagonal elements of $D^{(k)}$.

3. The kth compound of A' is $(A^{(k)})'$.

4. The kth compound of a Hermitian matrix is Hermitian.

5. The kth compound of a skew symmetric matrix is symmetric if k is even, skew symmetric if k is odd. State and prove the analogue for skew Hermitian matrices.

Next let us consider the products

$$A^{(k)}\,(\mathrm{adj}^{(k)}A) \quad \text{and} \quad (\mathrm{adj}^{(k)}A)A^{(k)}.$$

Every element in either product, in view of the way in which $\mathrm{adj}^{(k)}A$ has been defined, is a Laplacian expansion according to minors of order k (in the one case from rows of A, in the other from columns) and their cofactors, true or alien, in $|A|$. In the diagonal elements of the products the cofactors are the true, in the non-diagonal elements they are the alien cofactors. Hence in each product all non-diagonal elements are zero and all diagonal elements are equal to $|A|$, so that

$$A^{(k)} \operatorname{adj}^{(k)} A = \operatorname{adj}^{(k)} A \cdot A^{(k)} = |A| I, \qquad \cdot \quad (1)$$

all the matrices concerned being of order $n_{(k)}$. Taking determinants we have a theorem due to Cauchy,

$$|A^{(k)}| \, |\operatorname{adj}^{(k)} A| = |A|^{n(k)}. \qquad \cdot \qquad \cdot \qquad (2)$$

Compound Determinants, Sylvester's Theorem. Now $|A^{(k)}|$ and $|\operatorname{adj}^{(k)} A|$ are evidently both polynomials in the elements of A, of degrees $n_{(k)} k$ and $n_{(k)}(n-k)$ respectively ; and $|A|$ is itself a prime polynomial. It follows that both $|A^{(k)}|$ and $|\operatorname{adj}^{(k)} A|$ are powers of $|A|$, the possible numerical factor being seen to be 1, either by consideration of the term arising from the diagonal elements or merely by putting $A = \lambda I$. Further, since $|A|$ is of degree n and $|A^{(k)}|$ is of degree $n_{(k)} k$, while $|\operatorname{adj}^{(k)} A|$ is of degree $n_{(k)}(n-k)$, we have

$$|A^{(k)}| = |A|^{(n-1)(k-1)}, \quad |\operatorname{adj}^{(k)} A| = |A|^{(n-1)(k)}, \quad (3)$$

using $n_{(k)} k / n = (n-1)_{(k-1)}$, and $n_{(k)}(n-k)/n = (n-1)_{(k)}$.

The theorem in (3) on the value of the kth compound determinant was given by Sylvester in 1853. An important consequence is that if A is nonsingular so is $A^{(k)}$, while if A is singular so is $A^{(k)}$.

6. If $k = 1$, we have from (3) that $|\operatorname{adj} A| = |A|^{n-1}$, which is Cauchy's theorem on the value of the adjugate determinant.

7. If A is rectangular, of rank r, the $(r+1)^{th}$ and higher compounds of A are zero matrices, but $A^{(r)} \neq 0$.

8. $|A^{(n-k)}| = |A|^{(n-1)(n-k-1)} = |\operatorname{adj}^{(k)} A|$.

This is not surprising, in view of the fact (*cf.* **39**, Ex. **3**) that the $(n-k)^{th}$ compound and the kth adjugate compound have the same elements, namely, minors of $|A|$ of order $n-k$, but in the case of the adjugate compound they are transposed, written in the complete reverse of lexical order, and have cofactor sign imposed.

39. Binet–Cauchy Theorem on Product of Compound Matrices

We shall next prove a theorem of central importance, first given, in special cases, by Binet and Cauchy in 1812 : the kth compound of a product matrix AB is identically equal to the product of the kth compounds of A and B in that order.

Let A and B be rectangular matrices of orders $m \times n$, $n \times p$, and let us consider $A^{(k)}$, $B^{(k)}$, and $(AB)^{(k)}$.

Any element in AB is the product of a row vector of A into a column vector of B. More generally any submatrix of order $k \times k$ in AB is the product A_1B_1 of the submatrix A_1 consisting of a certain k rows of A, of course not necessarily consecutive, and the submatrix B_1 consisting of a certain k columns in B. This is most readily seen by supposing A partitioned into $\{A_1\ A_2\}$ according to those k rows and the remaining $m-k$, and B into $[B_1\ B_2]$ according to those k columns and the remaining $n-k$. For then we have

$$AB = \begin{bmatrix} A_1 \\ A_2 \end{bmatrix}[B_1\ B_2] = \begin{bmatrix} A_1B_1 & A_1B_2 \\ A_2B_1 & A_2B_2 \end{bmatrix}, \quad . \quad (1)$$

and A_1B_1 is the submatrix comprised by those k rows and k columns in AB. (*Cf.* also **11**.)

Now by the theorem on the determinant of a product of two rectangular matrices (**36** (4)) it follows that $|A_1B_1|$, which being a minor of AB of order k is an element of $(AB)^{(k)}$, is the sum of products of minors of order k drawn from k of the n columns of A_1 in all possible ways and minors of order k drawn from *corresponding* sets of k rows of B_1. But the minors from A_1 are all elements of a row of $A^{(k)}$ and the minors from B_1 are the corresponding elements of a column of $B^{(k)}$, and so the sum of products in question is an element of $A^{(k)}B^{(k)}$, by ordinary row-into-column multiplication. This is true for every element of $A^{(k)}B^{(k)}$, and the priority of elements in rows and columns is in strict accord with the lexical convention.

The conclusion is that the matrices $(AB)^{(k)}$ and $A^{(k)}B^{(k)}$ are element for element identical. Thus

$$(AB)^{(k)} = A^{(k)}B^{(k)}, \quad (ABC)^{(k)} = (AB)^{(k)}C^{(k)} = A^{(k)}B^{(k)}C^{(k)}, \quad (2)$$

and so step by step we derive the general result for the compound of the product of any finite number of matrices.

The elegance of this theorem is equalled only by the wealth of its applications. Since it rests on the theorem of Binet and Cauchy concerning the determinant of a product of rectangular matrices, and gives an extended matrix expression to that theorem, we shall call it the Binet-Cauchy theorem.

1. The more special theorem concerning the determinant of a product of rectangular matrices is that particular case of the Binet-Cauchy theorem in which A has k rows and B has k columns. Then $(AB)^{(k)}$ and $A^{(k)}B^{(k)}$ each consist of a single element.

The reader should take matrices of low order with literal and with numerical elements, and should familiarize himself with the import of these theorems.

2. Prove that $\qquad (\text{adj } A)^{(k)} = \lambda \text{ adj}^{(k)} A,$

where λ is a scalar, and show that λ is a certain power of $|A|$. (Premultiply both sides by $A^{(k)}$.)

3. Prove that if A is square of order $n \times n$ then

$$\text{adj } A = H(A')^{(n-1)}H^{-1},$$

where H is a matrix with all elements zero except those in the secondary diagonal at right angles to the principal diagonal, these elements being 1 and -1 alternately.

4. By taking the kth compound of $\bar{A}'A = I$ and applying the Binet-Cauchy theorem prove that the kth compound of a unitary matrix (24) is unitary.

Adjugate Binet-Cauchy Theorem. There is a theorem of the same nature as the Binet-Cauchy theorem, relating to the kth adjugate compound of a product. From **38** (1)

$$A^{(k)}B^{(k)}(\text{adj}^{(k)}B)(\text{adj}^{(k)}A) = |A||B|I, \quad . \quad (1)$$

and we have also

$$(AB)^{(k)}\text{adj}^{(k)}(AB) = |AB|I. \qquad . \qquad . \quad (2)$$

Since $|AB| = |A|\,|B|$ and $(AB)^{(k)} = A^{(k)}B^{(k)}$ we derive

$$\text{adj}^{(k)}(AB) = \text{adj}^{(k)}B \, . \, \text{adj}^{(k)}A, \qquad . \qquad . \quad (3)$$

and so step by step

$$\text{adj}^{(k)}(ABC) = \text{adj}^{(k)}C \, . \, \text{adj}^{(k)}(AB) = \text{adj}^{(k)}C \, . \, \text{adj}^{(k)}B \, . \, \text{adj}^{(k)}A, \quad (4)$$

and so on. The reversed order is to be noticed. The theorem expresses a set of polynomial identities in the elements a_{ij}, b_{ij} which persist when either or both of A and B are singular.

We may call the theorem the adjugate Binet-Cauchy theorem.

40. The Reciprocal of a Nonsingular Compound Matrix

If A is of order $n \times n$ and $|A| \neq 0$, we have $AA^{-1} = A^{-1}A = I$. Applying the Binet-Cauchy theorem we have

$$A^{(k)}(A^{-1})^{(k)} = (A^{-1})^{(k)}A^{(k)} = I, \qquad . \qquad . \quad (1)$$

where I is of order $n_{(k)}$. It follows that

$$(A^{(k)})^{-1} = (A^{-1})^{(k)}, \qquad . \qquad . \qquad . \quad (2)$$

in other words the reciprocal of the kth compound of A is identical with the kth compound of the reciprocal of A.

In the same way the adjugate Binet-Cauchy theorem yields the result that the reciprocal of the kth adjugate compound of A is identical with the kth adjugate compound of A^{-1}.

Add to these properties the almost intuitive ones that the kth compound of a Hermitian matrix is Hermitian, that the kth compound of A' is $(A^{(k)})'$, and that the kth compound of a unitary matrix is unitary (**39**, Ex. 4) and it becomes clear that there is a far-reaching parallelism between the properties of a matrix and those of its compounds or adjugate compounds. In fact, from any identity or equation involving products of matrices or reciprocal matrices we are able, by taking k^{th} compounds of every factor, to deduce a parallel identity or equation which is often more profound than the original one.

41. Rank of Matrix Expressed by Compound Matrices

Let A be of order $m \times n$ and of rank r. This implies that all minors formed from submatrices of order $(r+1) \times (r+1)$ vanish, while at least one minor of order r does not vanish. But this statement can be expressed more concisely in the form

$$A^{(r)} \neq 0, \ A^{(r+1)} = 0. \qquad . \qquad . \qquad (1)$$

Rank of Product Matrices. The Binet-Cauchy theorem now enables us to deduce at once certain important theorems on rank.

(i) Rank is invariant under multiplication by a non-singular matrix.

To prove this, consider A and HAK, where A is of rank r and $|H| \neq 0$, $|K| \neq 0$. Then $A^{(r)} \neq 0$, $A^{(r+1)} = 0$. Also

$$(HAK)^{(r+1)} = H^{(r+1)}A^{(r+1)}K^{(r+1)} = 0, \qquad . \qquad (2)$$

since $A^{(r+1)} = 0$. On the other hand,

$$(HAK)^{(r)} = H^{(r)}A^{(r)}K^{(r)} \neq 0, \qquad . \qquad (3)$$

otherwise we could divide out by the nonsingular matrices $H^{(r)}$ and $K^{(r)}$ and obtain $A^{(r)} = 0$, contrary to hypothesis.

Hence HAK is of rank r, as was to be proved. By putting $K = I$ we have the result for premultiplication alone ; and with $H = I$ for postmultiplication alone.

The following theorem concerning the rank of a product matrix is proved with equal ease :

(ii) The rank of a product matrix AB cannot exceed the rank of either factor A or B.

To prove this, suppose that the ranks of A and B are respectively r and s, where $r \leqslant s$. Then by the Binet-Cauchy theorem

$$(AB)^{(r+1)} = A^{(r+1)}B^{(r+1)} = 0, \qquad . \qquad . \quad (4)$$

since $A^{(r+1)} = 0$. Thus the rank of AB cannot exceed r, and a similar proof holds if $s \leqslant r$.

1. Take the matrices

$$\begin{bmatrix} 1 & . & . \\ . & 1 & . \\ . & . & 1 \end{bmatrix}, \begin{bmatrix} 1 & . & . \\ . & 1 & . \\ . & . & . \end{bmatrix}, \begin{bmatrix} . & 1 & . \\ . & . & 1 \\ . & . & . \end{bmatrix}, \begin{bmatrix} . & . & . \\ 1 & . & . \\ . & 1 & . \end{bmatrix}, \begin{bmatrix} . & . & 1 \\ . & . & . \\ . & . & . \end{bmatrix}.$$

form powers and products from them in various ways and note the rank, comparing it with the ranks of the factors.

Make experiments with other examples, choosing rectangular matrices of low order.

2. A matrix such as

$$U = \begin{bmatrix} . & 1 & . & . \\ . & . & 1 & . \\ . & . & . & 1 \\ . & . & . & . \end{bmatrix}$$

which satisfies an equation $U^s = 0$ is said to be *nilpotent*. A matrix M such that $M^2 = M$, and so $M^s = M$ for all positive integers s, is said to be *idempotent*.

Prove that the matrix $M = n^{-1}N$, where N is of order $n \times n$ and has every element $n_{ij} = 1$, is idempotent.

3. By the equivalent reduction HAK of 29 (7) prove that if A is of rank r, $A^{(k)}$ is of rank $r_{(k)}$.

42. Jacobi's Theorem on Minors of the Adjugate

Any minor of order r in the adjugate determinant of a square matrix A is, on expansion, a certain polynomial in the elements of A, fixed in form whether A is singular or not. To find an expression for this polynomial we may refer to the nonsingular case.

G

The kth compound of

yields
$$A \cdot \operatorname{adj} A = \operatorname{adj} A \cdot A = |A| I \qquad \qquad . \quad (1)$$

$$A^{(k)} (\operatorname{adj} A)^{(k)} = (\operatorname{adj} A)^{(k)} A^{(k)} = |A|^k I. \qquad . \quad (2)$$

But by taking the product of compound and adjugate compound we have also as in 38 (1)

$$A^{(k)} \operatorname{adj}^{(k)} A = \operatorname{adj}^{(k)} A \cdot A^{(k)} = |A| I. \qquad . \qquad . \quad (3)$$

Comparison of (2) and (3), A and so $A^{(k)}$ being non-singular, gives

$$(\operatorname{adj} A)^{(k)} = |A|^{k-1} \operatorname{adj}^{(k)} A. \qquad . \qquad . \quad (4)$$

Expressing this in words, we have the celebrated theorem of Jacobi (1834) concerning minors of the adjugate :

Any minor of order k in adj A is equal to the complementary signed minor in A', multiplied by $|A|^{k-1}$.

The relation thus established between the respective minors is the desired polynomial relation. It persists when A is singular, subject to the proviso that if $k = 1$ we must interpret $|A|^{k-1}$ as 1.

An alternative demonstration, the traditional one, does not depend on the Binet-Cauchy theorem, but proceeds as follows :

$$\begin{bmatrix} a_1 \; b_1 \; c_1 \; d_1 \\ a_2 \; b_2 \; c_2 \; d_2 \\ a_3 \; b_3 \; c_3 \; d_3 \\ a_4 \; b_4 \; c_4 \; d_4 \end{bmatrix} \begin{bmatrix} 1 \; |A_2| \; |A_3| \; . \\ . \; |B_2| \; |B_3| \; . \\ . \; |C_2| \; |C_3| \; 1 \\ . \; |D_2| \; |D_3| \; . \end{bmatrix} = \begin{bmatrix} a_1 \; . \; . \; c_1 \\ a_2 \; |A| \; . \; c_2 \\ a_3 \; . \; |A| \; c_3 \\ a_4 \; . \; . \; c_4 \end{bmatrix}. \quad (5)$$

The columns containing the specified minor of the adjugate, its elements denoted here by capital letters for cofactors, are left in place ; the remaining columns are deleted and a unit element is placed in each in the rows not containing the minor (as in 32, Ex.), so that the complementary submatrix of the minor is a unit submatrix ; and the matrix so constructed from adj A is premultiplied by A. On taking determinants we derive

$$|A| \times (\text{minor of adj } A) = |A|^k (\text{compl. signed minor in } A'), \quad (6)$$

from which Jacobi's theorem follows. The reader should examine the illustrative example, and should form other examples with A of order 5×5.

1. Construct adj A for a general matrix A of order 3×3 or 4×4 and verify Jacobi's theorem for various minors.

2. Construct several matrices of order 3×3 with simple numerical elements, and verify Jacobi's theorem for selected minors of the adjugate. Include some cases in which A is singular, and note specially the results when A has rank 2.

Minors of the Reciprocal Matrix. For nonsingular matrices Jacobi's theorem can be stated simply in respect of A^{-1}. For since $A^{-1} = |A|^{-1}$ adj A, any minor of order k in A^{-1} is equal to the corresponding minor in adj A multiplied by $|A|^{-k}$. Hence by (6) we have another formulation :

Any minor of order k in A^{-1} is equal to the complementary signed minor in A', multiplied by $|A|^{-1}$.

The phrase *any minor* may be taken to include $|A^{-1}|$ itself, the complementary minor in A' being then a minor of zero order, which by convention we take to be equal to unity.

The formulation just given might have been established at once by comparison of the two identities

$$A^{(k)}(A^{-1})^{(k)} = I \text{ and } A^{(k)}\text{adj}^{(k)}A = |A|I, \quad . \quad (7)$$

which yields the theorem in the form

$$(A^{-1})^{(k)} = |A|^{-1} \text{ adj}^{(k)}A. \quad . \quad (8)$$

Theorem of Bellavitis. If the cofactor $|A_{ij}|$ of a_{ij} in $|A|$ vanishes while the cofactor of $|a_{ij}\,a_{hk}|$ does not, then $|A|$ may be factorized into two rational factors, one linear in the elements of the ith row of A, the other linear in the elements of the jth column of A.

For there is no loss of generality in bringing a_{ij} and $|a_{ij}\,a_{hk}|$ into leading position. Then by Jacobi's theorem

$$\left| \begin{matrix} |A_{11}| & |A_{21}| \\ |A_{12}| & |A_{22}| \end{matrix} \right| = |A|\ |A_{12,\,12}|, \quad . \quad (9)$$

where $|A_{12,\,12}|$ is the cofactor of $|a_{11}\,a_{22}|$ in $|A|$. Hence since $|A_{11}| = 0$ we have

$$|A| = -|A_{12}|\ |A_{21}|/|A_{12,\,12}|. \qquad . \qquad . \quad (10)$$

Now $|A_{12}|$ can be expanded as a linear function of the elements of the first row of A, and $|A_{21}|$ as a linear function of elements of the first column. The theorem is thus proved.

3. Deduce that the bordered determinant

$$\begin{vmatrix} . & x' \\ y & A \end{vmatrix}, \quad \text{where } A \text{ is of rank } n-1$$

is the product of two linear functions, one linear in the elements of the row x', the other in the elements of the column y.

4. Deduce a sufficient condition that bilinear forms with square matrix (**31**, Ex. 1) or quadratic forms should be factorizable into rational linear factors. Construct some examples in two or three variables by taking bordered determinants with $|A|$ of rank $n-1$, and carry out the factorizing. For example:

$$-\begin{vmatrix} . & x_1 & x_2 \\ x_1 & 4 & 2 \\ x_2 & 2 & 1 \end{vmatrix} = x_1^2 - 4x_1 x_2 + 4x_2^2 = (x_1 - 2x_2)^2.$$

43. Franke's Theorem on Minors of a Compound Determinant

Jacobi's theorem can be extended at once to the case of any square matrices A and B satisfying a quasi-reciprocal relation $AB = \lambda I$, $\lambda \neq 0$. By the theorem on the rank of a product matrix both A and B must then be non-singular, so that $B = \lambda A^{-1}$. It follows from Jacobi's theorem that any minor of order k in B will be equal to the complementary signed minor in A' multiplied by $\lambda^k |A|^{-1}$.

For example, applying this to the identity

$$\text{adj}^{(k)} A \cdot A^{(k)} = |A|\,I \qquad . \qquad . \qquad . \quad (1)$$

and using Sylvester's theorem (**38**) on the value of the

adjugate compound determinant we deduce at once Franke's theorem :

Any minor of order h in the kth compound $A^{(k)}$ of a matrix A of order $n \times n$ is equal to the complementary signed minor in the kth adjugate compound transposed, multiplied by $|A|^{h-(n-1)(k)}$.

Sylvester's theorem is itself a special case of Franke's, bearing to it the same relation that Cauchy's theorem on the value of the adjugate determinant (**38**, Ex. 6) does to Jacobi's.

44. The Hybrid Compounds of Bazin and Reiss

Bazin in 1851 considered a compound determinant constructed as follows : Let A and B be two matrices each of order $n \times n$ and let a new matrix be constructed by replacing the jth row of A by the ith row of B in all possible ways, the n^2 determinants of the resulting matrices being taken as $(i, j)^{th}$ elements of the matrix so derived. This matrix is Bazin's matrix. We shall call it the first *hybrid compound* of A and B taken in that order, and we shall denote it by $(A \, ; B)$.

Now $(A \, ; B)$ is in fact the same as B adj A. For just as each element in A adj A is an expansion, namely, by elements of the ith row of A and cofactors of corresponding elements of the jth row, so, with B substituted for A in the first factor, each element in B adj A is an expansion, by elements of the ith row of B and cofactors of corresponding elements of the jth row of A. But such expansions (**21** (1)) are those of determinants obtained by substituting the ith row of B for the jth row of A ; and so Bazin's matrix is B adj A, as stated.

Taking determinants we have $|(A \, ; B)| = |B| \, |A|^{n-1}$, Bazin's theorem.

1. Put $B = I$. Then $(A \, ; I) = \text{adj} A$, and Bazin's theorem becomes Cauchy's theorem on the adjugate.

2. What is the relation of $(I \, ; B)$ to B ? Write out an example of order 3×3.

Reiss in 1867 (and Picquet in 1878) considered from the standpoint of determinants a more general matrix obtained from A and B in a manner resembling Bazin's, except that not one but k rows of B were substituted for k rows of A in all possible ways, the determinants of the resulting matrices being taken as elements of a matrix, arranged in rows and columns according to lexical order of the substituted rows. It is easy to see that this Reissian matrix, of order $n_{(k)} \times n_{(k)}$, is the same as $B^{(k)}\operatorname{adj}^{(k)}A$. For just as the elements of $A^{(k)}\operatorname{adj}^{(k)}A$ are Laplacian expansions (**33**) by minors from k rows of A and cofactors of corresponding minors from either the same or a different set of k rows of A, so the elements in $B^{(k)}\operatorname{adj}^{(k)}A$ are Laplacian expansions in which the first set of minors from k rows of A just mentioned is replaced by the corresponding set in B. But these are all Laplacian expansions of determinants obtained by substituting k rows of B for k rows of A; and the principle of lexical order is observed. Hence the matrix of Reiss is $B^{(k)}\operatorname{adj}^{(k)}A$, as stated.

Taking determinants we have a theorem of Reiss :

The determinant of Reiss is equal to

$$|B|^{(n-1)(k-1)} \; |A|^{(n-1)(k)} \qquad . \qquad . \quad (2)$$

3. Bazin's is the case $k = 1$. The case $B = I$ gives Sylvester's theorem on the kth adjugate compound determinant.

4. What is the relation of the kth Reissian of I and B to B ?

5. Show that the coefficient of $\lambda^{n-k}\mu^k$ in $|\lambda A + \mu B|$ is the trace of $B^{(k)} \operatorname{adj}^{(k)}A$, the kth Reissian hybrid.

From A and B another Bazin matrix arises, $A \operatorname{adj} B$. Now

$$A \operatorname{adj} B . B \operatorname{adj} A = |AB|I, \qquad . \qquad . \quad (3)$$

and so these dual hybrid compounds are quasi-reciprocal. It follows from Jacobi's theorem that any minor of order h in $A \operatorname{adj} B$ is equal to the complementary signed minor in $(B \operatorname{adj} A)'$, multiplied by a certain factor, which will evidently be a product of powers of $|B|$ and $|A|$. We leave the reader to prove, by a consideration of dimensions, or simply by putting $B = I$, $A = I$ in turn, that these powers are respectively the $(h-1)^{th}$ and the $(h-n+1)^{th}$.

In exactly the same way, since

$$A^{(k)}\text{adj}^{(k)}B \ . \ B^{(k)}\text{adj}^{(k)}A = |AB|I, \qquad . \quad (4)$$

we see that any minor of order h in the kth dual Reissian hybrid is equal to the complementary signed minor in the transposed k^{th} Reissian, multiplied by certain powers of $|B|$ and $|A|$. Once again these powers can be found easily, either by consideration of dimensions or by putting $B = I$, $A = I$ in turn and referring to the special case, namely, Franke's theorem ; and so we find the multiplying factor to be

$$|B|^{h-(n-1)(k-1)} \quad |A|^{h-(n-1)(k)}. \qquad . \qquad . \quad (5)$$

This last theorem, which is also due to Reiss, is the most general of the theorems in this class ; the other theorems of Reiss himself, and those of Bazin, Franke, Sylvester, Jacobi and Cauchy, are successive special cases of it.

6. Since $\qquad (B \text{ adj } A)^{(k)} = |A|^{k-1}B^{(k)}\text{adj}^{(k)}A,$

the kth Reissian hybrid matrix is simply, apart from a scalar factor, the kth compound of Bazin's matrix ; which suggests alternative derivations for the theorems proved above.

45. Complementary Identities : Extensional Identities

The reader will by now have realised that the appearance of generality of several of the above theorems is deceptive, and that the really useful theorems are the two on which all the others rest, the Binet-Cauchy theorem and Jacobi's theorem. We shall now see how Jacobi's theorem furnishes the means of deriving valuable dual identities and extended identities in determinants.

Let determinants and their minors be indicated by diagonal elements in a single-suffix notation, and let cofactors be denoted by corresponding Greek letters ; so that for example the transposed adjugate determinant of

$|a_1b_2c_3d_4|$ will be $|a_1\beta_2\gamma_3\delta_4|$. Now consider any homogeneous polynomial identity involving minors of a determinant, of any orders. Then such an identity, being general, is true also for the transposed adjugate, and so may be written with Greek letters. Next, by Jacobi's theorem we may replace every minor of the adjugate by its complementary signed minor in A', multiplied by some power of $|A|$. In such replacements $|A|$ itself may be regarded as the complementary minor of a minor of order zero, the latter being taken by convention to be equal to 1. This substitution of minors will thus give us another identity, which may or may not be different in form from the first identity. In any case the original identity and this derived one stand in a dual relation, and may be called *dual* or *complementary* identities with respect to A.

1. Consider the identity $|a_1b_2c_3| = ||a_1||b_2||c_3||$, where the notation on the right indicates that we are to view the elements as minors of the first order. Now the complementary signed minor of a_1, for example, is $|b_2c_3|$, and so on, and applying Jacobi's theorem we derive the dual identity

$$|a_1b_2c_3|^2 = |\text{adj } A|,$$

an example of Cauchy's theorem on the adjugate. The reader will easily extend this to the case of general order.

2. Consider a determinant $|A| = |a_1b_2c_3d_4|$; partition it according to its 1st row, its 2nd row and its last two rows, and then expand it as a general Laplacian expansion

$$|a_1b_2c_3d_4| = |a_1||b_2||c_3d_4| - |a_1||c_2||b_3d_4| + \cdots$$

of 12 terms. Taking complementary minors, and making some sign changes in rows and columns, we obtain

$$|a_1b_2c_3d_4|^2 = |a_1b_2||a_1c_3d_4||b_2c_3d_4| - |a_1c_2||a_1b_3d_4||b_2c_3d_4| + \cdots,$$

an identity far from obvious.

But the consequences of Jacobi's theorem are wider still. Any minor such as $|a_1b_2|$ is not merely a minor of $|a_1b_2c_3d_4|$, it is also a minor of $|a_1b_2c_3d_4e_5f_6g_7|$, where $(e_5f_6g_7)$, which we shall term an *extension* of $|a_1b_2c_3d_4|$, denotes the presence of added rows and columns. Now, given any identity, we may as before write down its complementary. We can then take the complementary of this, not with respect to the original determinant, which would merely give us back our first identity, but with respect to the enlarged or extended determinant such as $|a_1b_2c_3d_4e_5f_6g_7|$. When this is done, every minor in the original identity, including minors of order zero and the determinant itself, reappears with a tail or extension added to it. For example the complementary of

$$|a_1b_2| = |\,|a_1|\,|b_2|\,| \text{ is } |a_1b_2| = \left|\begin{matrix} |b_2| & -|a_2| \\ -|b_1| & |a_1| \end{matrix}\right|, \quad (1)$$

a very trivial result, but the complementary of this with respect to $|a_1b_2c_3d_4|$, for example, yields

$$|a_1b_2c_3d_4|\,|c_3d_4| = \left|\begin{matrix} |a_1c_3d_4| & |b_1c_3d_4| \\ |a_2c_3d_4| & |b_2c_3d_4| \end{matrix}\right|, \quad . \quad (2)$$

which is by no means trivial and is indeed one of the identities on which (**21** (6)) the method of evaluating determinants by pivotal condensation was based. That the identity is extensional in nature is seen by the extension (c_3d_4) which is apparent in every indicated minor. The reader will generalize the result.

The rule so derived for the existence or the construction of extensional identities may be expressed as follows :

If any homogeneous polynomial identity involving the minors of a determinant be written in terms of the diagonal elements of the minors concerned, then a new identity can be obtained by extending or elongating all such diagonals by further elements, such as $(e_5f_6g_7)$, provided that homogeneity be maintained by the insertion of powers of the determinant of the extension, for example $|e_5f_6g_7|$. Such a determinant is to be regarded as the extension of a minor of order zero.

3. The extensional of

$$|a_1b_2c_3| = |a_1||b_2c_3| - |a_2||b_1c_3| + |a_3||b_1c_2|$$

by the extension (d_4e_5) is

$$|a_1b_2c_3d_4e_5||d_4e_5| = |a_1d_4e_5||b_2c_3d_4e_5| - |a_2d_4e_5||b_1c_3d_4e_5| \\ + |a_3d_4e_5||b_1c_2d_4e_5|.$$

4. The extensional of Sylvester's theorem

$$||a_1b_2||a_1c_3||a_1d_4||b_2c_3||b_2d_4||c_3d_4|| = |a_1b_2c_3d_4|^3,$$

where we indicate on the left the 2nd compound of $|a_1b_2c_3d_4|$, by the extension (e_5f_6) yields

$$||a_1b_2e_5f_6||a_1c_3e_5f_6|\cdots|c_3d_4e_5f_6|| = |a_1b_2c_3d_4e_5f_6|^3|e_5f_6|^3,$$

a theorem which, stated for general orders, is often called Sylvester's theorem. It is in fact, as we see, a simple extensional of Sylvester's theorem on the kth compound determinant. This extensional theorem has a deceptive resemblance to Reiss' theorem on the determinant of a hybrid compound (**44** (2)), which is not an extensional theorem.

5. A theorem of D'Ovidio (1877) is to the effect that a compound determinant having for elements all minors of order k not entirely contained in the last h rows and h columns of A, where A is of order $n \times n$, has the value

$$|A_1|^{(h-1)(k-1)}|A|^{(n-1)(k-1)-(h-1)(k-1)},$$

where A_1 denotes the submatrix composed of the first $n-h$ rows and $n-h$ columns of A.

The matrix of D'Ovidio's compound is a submatrix of $A^{(k)}$, of order $n_{(k)}-h_{(k)}$. The complementary submatrix in the transposed k^{th} adjugate compound of A is easily perceived to be the extensional of the transposed k^{th} adjugate compound of A_2, where A_2 is complementary to A_1 in A, by the extension (A_1'). Thus the theorem is really an extensional of Franke's theorem (**43**) in a rather disguised enunciation, and the exponents are easily found.

Let us take the opportunity of remarking that the extensional nature of an identity may sometimes be concealed by a permutation of rows or columns in a minor, deranging its diagonal representation. For example, if $|a_1b_2c_3d_4|$ be written $-|a_1c_2b_3d_4|$, the presence of the extension (c_3d_4) may escape notice.

Vanishing Aggregates of Products.* Many identities found by early writers, in the form of vanishing aggregates of products of determinants, may be recognised

* The paragraph standing at this place in earlier editions was incorrect.

as extensionals of simpler basic identities. For example, there is a class of identities given in 1809 by Monge, of which the following is typical :

$$|a_1b_2c_3||c_1d_2e_3| + |b_1c_2d_3||a_1c_2e_3| - |a_1c_2d_3||b_1c_2e_3| = 0.$$

If it is rearranged in the shape

$$|a_1b_2c_3||d_1e_2c_3| - |a_1d_2c_3||b_1e_2c_3| + |a_1e_2c_3||b_1d_2c_3| = 0$$

we recognise the extensional, by (c_3), of the identity

$$|a_1b_2||d_1e_2| - |a_1d_2||b_1e_2| + |a_1e_2||b_1d_2| = 0$$

readily proved by Laplacian expansion of $|a_1b_2c_1d_2|$.

Again, among identities given in 1819 by Desnanot there appears

$$|a_1b_2||a_1b_3c_4| - |a_1b_3||a_1b_2c_4| + |a_1b_4||a_1b_2c_3| = 0.$$

This is visibly the extensional of

$$b_2|b_3c_4| - b_3|b_2c_4| + b_4|b_2c_3| = 0,$$

the obvious expansion of a vanishing determinant.

6. By considering $|a_1b_2c_3d_1e_2f_3|$, expanded first according to minors from the last two columns, and then each cofactor by its first row, establish the vanishing of

$$|a_1b_2c_3||d_1e_2f_3| - |a_1b_2d_3||c_1e_2f_3| + |a_1c_2d_3||b_1e_2f_3| - |b_1c_2d_3||a_1e_2f_3|.$$

A prolific source of identities is here indicated.

46. Schweinsian Expansions of Determinant Quotients

Quotients of determinants of the type $|h_1b_2c_3d_4|/|a_1b_2c_3d_4|$ and $|a_1b_2c_3d_4|/|b_2c_3d_4|$ are common in applications of determinants, the first of them occurring (**22** (4)) in the solution of simultaneous equations. In the first quotient the numerator differs from the denominator in the first column only. The second quotient is of the form $|A|/|A_{11}|$.

In the general case the determinants in numerator and denominator are of the nth or the nth and $(n-1)^{th}$, order.

Such quotients may be expanded in series of a type first given by Schweins in 1825. The procedure will be sufficiently illustrated by quotients with numerators of the 4th order.

(i) Let $|h_1 b_2 c_3 d_4|/|a_1 b_2 c_3 d_4| = q_4$, $|h_1 b_2 c_3|/|a_1 b_2 c_3| = q_3$.

Then $q_4 - q_3 = \left\| \begin{matrix} |h_1 b_2 c_3 d_4| & |a_1 b_2 c_3 d_4| \\ |h_1 b_2 c_3| & |a_1 b_2 c_3| \end{matrix} \right\| / |a_1 b_2 c_3| \; |a_1 b_2 c_3 d_4|.$ (1)

The numerator of (1) is clearly an extensional, by the extension $(b_2 c_3)$, of

$$\left| \begin{matrix} |h_1 d_4| & |a_1 d_4| \\ h_1 & a_1 \end{matrix} \right| = |h_1 a_4| d_1,$$

and so is equal to

$$|h_1 a_4 b_2 c_3| \; |d_1 b_2 c_3| = |h_1 a_2 b_3 c_4| \; |b_1 c_2 d_3|,$$

by an even number of interchanges of rows and columns. So

$$q_4 - q_3 = |h_1 a_2 b_3 c_4| \; |b_1 c_2 d_3| / |a_1 b_2 c_3| \; |a_1 b_2 c_3 d_4|. \qquad (2)$$

In the same way

$$q_3 - q_2 = |h_1 a_2 b_3| \; |b_1 c_2| / |a_1 b_2| \; |a_1 b_2 c_3|, \qquad (3)$$
$$q_2 - q_1 = |h_1 a_2| |b_1 / a_1| a_1 b_2|, \qquad (4)$$
$$q_1 = h_1 / a_1. \qquad (5)$$

Adding now the results (2), (3), (4), (5) we have the first Schweinsian expansion

$$|h_1 b_2 c_3 d_4|/|a_1 b_2 c_3 d_4| = h_1 / a_1 + |h_1 a_2| |b_1 / a_1| a_1 b_2|$$
$$+ |h_1 a_2 b_3| \; |b_1 c_2| / |a_1 b_2| \; |a_1 b_2 c_3|$$
$$+ |h_1 a_2 b_3 c_4| \; |b_1 c_2 d_3| / |a_1 b_2 c_3| \; |a_1 b_2 c_3 d_4|, \qquad (6)$$

The general theorem of this kind, for which an exactly similar proof may be constructed, follows the above pattern.

(ii) Let $|a_1 b_2 c_3 d_4|/|b_2 c_3 d_4| = p_4$, $|a_1 b_2 c_3|/|b_2 c_3| = p_3$.

Then $p_4 - p_3 = \left| \begin{matrix} |a_1 b_2 c_3 d_4| & |b_2 c_3 d_4| \\ |a_1 b_2 c_3| & |b_2 c_3| \end{matrix} \right| / |b_2 c_3| \; |b_2 c_3 d_4|. \qquad (7)$

Here again the numerator is an extensional, by (b_2c_3), of

$$\begin{vmatrix} |a_1d_4| & d_4 \\ a_1 & 1 \end{vmatrix} = -a_4d_1,$$

and so is equal to

$$-|a_4b_2c_3|\ |d_1b_2c_3| = -|a_2b_3c_4|\ |b_1c_2d_3|,$$

again by an even number of interchanges of rows and columns. So

$$p_4-p_3 = -|a_2b_3c_4|\ |b_1c_2d_3|/|b_2c_3|\ |b_2c_3d_4|, \qquad . \quad (8)$$

and by adding this to the similar results for p_3-p_2, p_2-p_1, and $p_1 = a_1$, we derive the second Schweinsian expansion

$$|a_1b_2c_3d_4|/|b_2c_3d_4| = a_1-a_2b_1/b_2-|a_2b_3|\ |b_1c_2|/b_2|b_2c_3|$$
$$-|a_2b_3c_4|\ |b_1c_2d_3|/|b_2c_3|\ |b_2c_3d_4|. \qquad . \quad (9)$$

The form of the general result, which is proved in the same way, is clearly seen. The Schweinsian expansions (6) and (9) have what may be called the " extra shutter " property, namely, that if a term be added similar to the last term but with every minor suitably extended, the sum of the series then yields the next higher quotient in the sequence. We have assumed throughout that all minors in the denominators are non-vanishing.

1. If $|a_1b_2c_3d_4| = 0$, $|b_2c_3d_4| \neq 0$, then

$$a_1 = a_2b_1/b_2+|a_2b_3||b_1c_2|/b_2|b_2c_3|+|a_2b_3c_4||b_1c_2d_3|/|b_2c_3||(b_2c_3d_4), \quad (10)$$

and so in general.

2. If u_x is a cubic polynomial $u_x = c_0+c_1x+c_2x^2+c_3x^3$, and if

$$\int_0^1 u_x dx = m_0, \quad \int_0^1 x u_x dx = m_1, \quad \int_0^1 x^2 u_x dx = m_2, \quad \int_0^1 x^3 u_x dx = m_3,$$

we may write down corresponding to these five conditions five homogeneous linear equations in 1, c_0, c_1, c_2, c_3. The condition of consistency is that the determinant of the system, namely,

$$\begin{vmatrix} u_x & 1 & x & x^2 & x^3 \\ m_0 & 1 & \frac{1}{2} & \frac{1}{3} & \frac{1}{4} \\ m_1 & \frac{1}{2} & \frac{1}{3} & \frac{1}{4} & \frac{1}{5} \\ m_2 & \frac{1}{3} & \frac{1}{4} & \frac{1}{5} & \frac{1}{6} \\ m_3 & \frac{1}{4} & \frac{1}{5} & \frac{1}{6} & \frac{1}{7} \end{vmatrix} = 0.$$

If the reader now regards this determinant as $|a_1 b_2 c_3 d_4 e_5|$ and applies (10), u_x will be obtained in terms of polynomials

$$1, \begin{vmatrix} 1 & x \\ 1 & \frac{1}{2} \end{vmatrix}, \begin{vmatrix} 1 & x & x^2 \\ 1 & \frac{1}{2} & \frac{1}{3} \\ \frac{1}{2} & \frac{1}{3} & \frac{1}{4} \end{vmatrix}, \begin{vmatrix} 1 & x & x^2 & x^3 \\ 1 & \frac{1}{2} & \frac{1}{3} & \frac{1}{4} \\ \frac{1}{2} & \frac{1}{3} & \frac{1}{4} & \frac{1}{5} \\ \frac{1}{3} & \frac{1}{4} & \frac{1}{5} & \frac{1}{6} \end{vmatrix}.$$

If these polynomials be denoted by $p_0(x)$, $p_1(x)$, $p_2(x)$, $p_3(x)$, the reader will easily prove that they satisfy the condition

$$\int_0^1 x^s p_r(x)\,dx = 0, \quad s < r.$$

3. Let the polynomial $f(x) = c_0 + c_1 x + c_2 x^2$ be given by three values $f(a_0)$, $f(a_1)$, $f(a_2)$, where a_0, a_1, a_2 are distinct. Then the consistency of the four equations for $f(x)$, $f(a_0)$, $f(a_1)$, $f(a_2)$ in terms of 1, c_0, c_1, c_2 is expressed by the vanishing of a determinant of the 4th order. Write it down and, by applying (10) to it, obtain an expansion for $f(x)$ in terms of $f(a_0)$, $f(a_1)$, $f(a_2)$. (The expansion is Newton's interpolation formula of divided differences.)

4. Expand the quadratic form

$$- \begin{vmatrix} . & x_1 & x_2 & x_3 \\ x_1 & a_{11} & a_{12} & a_{13} \\ x_2 & a_{21} & a_{22} & a_{23} \\ x_3 & a_{31} & a_{32} & a_{33} \end{vmatrix} \Bigg/ \begin{vmatrix} a_{11} & a_{12} & a_{13} \\ a_{21} & a_{22} & a_{23} \\ a_{31} & a_{32} & a_{33} \end{vmatrix}$$

where the bordered matrix A in the numerator is symmetric, by the second Schweinsian expansion. Note that the result gives the quadratic form in terms of squares of certain linear forms in x_1, x_2, x_3.

5. Note that $\operatorname{sp} A^{(k)} = \operatorname{sp}_k A$. (See **37**, (3).)

6. The Binet-Cauchy theorem applied to $Q^{-1}AQ = \Lambda$ (**37**, Ex. 7) gives $(Q^{(k)})^{-1} A^{(k)} Q^{(k)} = \Lambda^{(k)}$.

This is clearly a reduction of $A^{(k)}$ to canonical form $\Lambda^{(k)}$, a diagonal matrix (**38**, Ex. 2) with k-ary products of the latent roots of A in its diagonal. These k-ary products are therefore the latent roots of $A^{(k)}$. The factor $Q^{(k)}$ also shows that the latent vectors of $A^{(k)}$ are k^{th} compounds of the $n_{(k)}$ matrices formed by juxtaposing k latent column vectors of A.

CHAPTER VI

SPECIAL DETERMINANTS : ALTERNANT, PERSYMMETRIC, BIGRADIENT AND CENTROSYMMETRIC, JACOBIANS, HESSIANS, WRONSKIANS

IN the present chapter we survey a number of special types of determinant which occur persistently in the solution of certain general mathematical problems.

47. Alternant Matrices and Determinants

A *symmetric function* is a function of several variables which remains unaltered when any two of the variables are interchanged. An *alternating function* is one which is merely altered in *sign* by such an interchange. It is evident at once that the reciprocal of a symmetric function, the product of any number of symmetric functions and the product of any *even* number of alternating functions are symmetric functions ; while the reciprocal of an alternating function, the product of a symmetric and an alternating function and the product of any *odd* number of alternating functions are alternating functions. Constants, being independent of the variables, are to be regarded as symmetric functions of zero degree.

1. Examples of symmetric functions : x_1+x_2, $x_1+x_2+x_3$, $x_1x_2+x_2x_3+x_3x_1$, $x_1x_2x_3$, $x_1^2+x_2^2+x_3^2$, $\sin(x_1+x_2)$, $\cos(x_1+x_2)$, $\cosh(x_1+x_2)$, $(x_1-x_2)^2+(x_2-x_3)^2+(x_3-x_1)^2$.

2. Alternating functions : x_1-x_2, $(x_1-x_2)(x_2-x_3)(x_3-x_1)$, $(x_1-x_2)^3$, $\sin(x_1-x_2)$, $\sinh(x_1-x_2)$, $(x_1+x_2)/(x_1-x_2)$.

A function $f(x)$ is often expressed linearly in terms of a

111

linearly independent set of basic functions $p_0(x)$, $p_1(x)$, ..., $p_{n-1}(x)$ as follows:

$$f(x) = c_0 p_0(x) + c_1 p_1(x) + \cdots + c_{n-1} p_{n-1}(x). \qquad (1)$$

For example the functions $p_j(x)$ may be powers x^j, or may be trigonometric functions $\cos jx$ or $\sin jx$. Each special problem has its own appropriate basic functions $p_j(x)$.

When x takes the m values x_1, x_2, ..., x_m let the matrix of order $m \times n$ which has $p_{j-1}(x_i)$ for its $(i, j)^{th}$ element be denoted by P. Then the set of m values $f(x_i)$ may be regarded as a column vector and in view of (1) may be written

$$f = Pc, \qquad \cdot \qquad \cdot \qquad \cdot \qquad (2)$$

where c is the column vector $\{c_0\ c_1\ \cdots\ c_{n-1}\}$.

Such a matrix $P = [p_{j-1}(x_i)]$ of functional values is called an *alternant matrix*, and its determinant $|P|$, when P is square, is called an *alternant*. It is so called because to interchange two of the variables, such as x_i and x_k, is to interchange the ith and k^{th} rows of $|P|$, and therefore to change its sign. Hence $|P|$ is an alternating function in x_1, x_2, ..., x_m.

3. $|x_1^0\ x_2^1\ x_3^2| \equiv \begin{vmatrix} 1 & x_1 & x_1^2 \\ 1 & x_2 & x_2^2 \\ 1 & x_3 & x_3^2 \end{vmatrix}$ and $|x_1^0\ x_2^q\ x_3^r| \equiv \begin{vmatrix} 1 & x_1^q & x_1^r \\ 1 & x_2^q & x_2^r \\ 1 & x_3^q & x_3^r \end{vmatrix}$

are alternants, the basic functions $p_j(x)$ being the powers $1, x, x^2$ and $1, x^q, x^r$ respectively.

4. Write out in full the alternants briefly denoted, after the manner of Ex. 3, by $|x_1^0 \cos x_2 \sin x_3|$, $|x_1^2 \cot x_2\ x_3^3|$.

Polynomial Alternants. The Simple Alternant.

The alternant $|x_1^0\ x_2^1\ x_3^2\ \cdots\ x_n^{n-1}|$ is the product (**19,** Ex. 11) of the $\frac{1}{2}n(n-1)$ differences of x_1, x_2, ..., x_n taken in pairs and written in reversed order. We shall denote it by $|A(012\ldots\overline{n-1})|$, leaving the arguments x_i to be understood, and we shall call it the *simple* alternant.

5. Prove by operations on columns that if instead of the powers $1, x, x^2, \ldots, x^{n-1}$ we take as basic functions polynomials of degree $0, 1, 2, \ldots, n-1$ having $1, x, x^2, \ldots, x^{n-1}$ respectively as terms of highest degree, then the resulting alternant is equal to the simple alternant $|A(012...\overline{n-1})|$.

Try some numerical examples of low order, taking polynomials such as 1, x, $x(x-1)$, $x(x-1)(x-2)$ for basic functions.

6. Prove that if the basic functions are polynomials of degree, $0, 1, 2, ..., n-1$ with respective terms of highest degree d_0, d_1x, d_2x^2, ..., $d_{n-1}x^{n-1}$, then the alternant is equal to $d_0d_1d_2...d_{n-1}|A(012...n-1)|$.

The more general polynomial alternant based on n powers 1, x^q, x^r, ..., x^t, $0<q<r<...<t$, may be denoted by $|A(0qr...t)|$. There is no loss of generality in taking the first index to be 0, since if it were p we could cancel x_1^p, x_2^p, ..., x_n^p from the respective rows of the alternant.

The alternants $|A(0qr...t)|$ and $|A(012...\overline{n-1})|$ are easily seen to be homogeneous alternating polynomials in x_1, x_2, ..., x_n. Further, $|A(0qr...t)|$ vanishes when we put any $x_i = x_k$, because the ith and k^{th} rows then become identical. Hence by the remainder theorem it must contain the difference-product $|A(012...\overline{n-1})|$ as a factor, so that the quotient of alternants

$$|A(0qr...t)|/|A(012...\overline{n-1})| \qquad . \qquad . \qquad . \quad (3)$$

must be a homogeneous symmetric polynomial. It is called a *bialternant*, and is fundamental in the theory of symmetric polynomials, in algebraic invariants and in more general domains of algebra. Determinantal expressions for it will be given in **49**.

7. Evaluate and examine the bialternants $|A(024)/|A(012)|$, $|A(014)|/|A(012)|$ and $|A(023)|/|A(012)|$ in x_1, x_2, x_3.

48. Elementary and Complete Homogeneous Symmetric Functions

The sum of the $n_{(j)}$ products of x_1, x_2, ..., x_n taken j at a time, without repetition of any x_i in a product, is called the *elementary symmetric function* of degree j in the arguments x_1, x_2, ..., x_n, and is often denoted by a_j, the arguments being tacit. We shall use a_j in this sense as far as 50. The sum of the $(n+j-1)_{(j)}$ products of the $x_1, x_2, ..., x_n$, taken j at a time and with unrestricted repetition of any x_i in a product (so that powers up to the jth may appear), is called the *complete homogeneous symmetric function* of degree j, and will be denoted by h_j. It is sometimes called the *aleph function* of degree j, a name used by Wronski in 1812. We note that $a_j = 0, j > n$.

1. The following are elementary symmetric functions in the arguments concerned : $x_1 + x_2 + x_3,\ x_1 x_2 + x_1 x_3 + x_2 x_3,\ x_1 x_2 x_3$.

2. The following are the corresponding h_j : $x_1 + x_2 + x_3$, $x_1^2 + x_2^2 + x_3^2 + x_1 x_2 + x_1 x_3 + x_2 x_3,\ \Sigma x_i^3 + \Sigma x_i^2 x_k + x_1 x_2 x_3$.

It is easily proved that a_j is the coefficient of $(-)^j t^j$ in the expanded form of the generating function

$$(1 - x_1 t)(1 - x_2 t)\ ...\ (1 - x_n t),\qquad\qquad . \qquad . \quad (1)$$

while h_j is the coefficient of t^j in the expansion of

$$(1 + x_1 t + x_1^2 t^2 + ...)(1 + x_2 t + x_2^2 t^2 + ...)...(1 + x_n t + x_n^2 t^2 + ...),\ (2)$$

the terms in t^2, t^3, ... in the factors of (2) taking account of the repetitions of the x_i. Now the generating function (2) is visibly the reciprocal of (1), and its expansion is a series which converges if t^{-1} is less in absolute value than any of $x_1, x_2, ..., x_n$. Consequently by multiplication of (1) and (2)

$$1 = (1 - a_1 t + a_2 t^2 - a_3 t^3 + ... + (-)^n a_n t^n)(1 + h_1 t + h_2 t^2 + ...),\ (3)$$

from which by comparing coefficients of powers of t we have

$$a_1 h_0 - a_0 h_1 = 0,$$
$$a_2 h_0 - a_1 h_1 + a_0 h_2 = 0,$$
$$\cdots\cdots\cdots\cdots\cdots \qquad\qquad . \qquad (4)$$
$$a_k h_0 - a_{k-1} h_1 + ... + (-)^k a_0 h_k = 0,$$

where by convention $a_0 = h_0 = 1$. These relations, which are evidently unaltered when the rôles of the a_j and h_j are interchanged, are called Wronski's relations. Let us observe that they are expressed completely in matrix notation by the statement that

$$A \equiv \begin{bmatrix} a_0 & & \\ -a_1 & a_0 & \\ a_2 & -a_1 & a_0 \\ \cdots \cdots \cdots \\ \cdots \cdots \cdots \end{bmatrix} \text{ and } H \equiv \begin{bmatrix} h_0 & & \\ h_1 & h_0 & \\ h_2 & h_1 & h_0 \\ \cdots \cdots \cdots \\ \cdots \cdots \cdots \end{bmatrix} \quad (5)$$

are reciprocal matrices, as the reader should verify by proving that $AH = HA = I$. Indeed since the determinants $|A|$ and $|H|$ are both equal to 1, the matrices A and H are mutually adjugate as well as reciprocal.

49. Bialternant Symmetric Functions of Jacobi

Let us take three arguments x_1, x_2, x_3, or, for greater ease in printing, a, β, γ and let us consider $h_r(a, \beta, \gamma)$. The terms in this symmetric function may be divided into two groups, those containing a and those not containing a. The latter set clearly gives $h_r(\beta, \gamma)$; and slight consideration will show that the former set gives $ah_{r-1}(a, \beta, \gamma)$.

Hence
$$h_r(a, \beta, \gamma) = h_r(\beta, \gamma) + ah_{r-1}(a, \beta, \gamma)$$
$$= h_r(a, \gamma) + \beta h_{r-1}(a, \beta, \gamma), \quad \cdot \quad (1)$$

whence it follows, on subtraction, that the divided difference
$$\{h_r(\beta, \gamma) - h_r(a, \gamma)\}/(\beta - a) = h_{r-1}(a, \beta, \gamma). \quad \cdot \quad (2)$$

A corresponding identity can be proved in the same way for any number of arguments. Now let us take the alternant

$$|A(0rst)| \equiv \begin{vmatrix} 1 & a^r & a^s & a^t \\ 1 & \beta^r & \beta^s & \beta^t \\ 1 & \gamma^r & \gamma^s & \gamma^t \\ 1 & \delta^r & \delta^s & \delta^t \end{vmatrix} \quad \cdot \quad \cdot \quad \cdot \quad (3)$$

and subject it to the following operations : $\text{row}_2 - \text{row}_1$ and remove the factor $\beta - a$; $\text{row}_3 - \text{row}_1$ and remove $\gamma - a$; $\text{row}_4 - \text{row}_1$ and remove $\delta - a$; then $\text{row}_3 - \text{row}_2$

and remove $\gamma-\beta$, $\text{row}_4-\text{row}_2$ and remove $\delta-\beta$; and so on. By repeated use of (2), with the evident fact at the first stage of $h_r(a)=a^r$ and so on, we thus arrive at

$$\begin{vmatrix} 1 & h_r(a) & h_s(a) & h_t(a) \\ \cdot & h_{r-1}(a,\beta) & h_{s-1}(a,\beta) & h_{t-1}(a,\beta) \\ \cdot & h_{r-2}(a,\beta,\gamma) & h_{s-2}(a,\beta,\gamma) & h_{t-2}(a,\beta,\gamma) \\ \cdot & h_{r-3}(a,\beta,\gamma,\delta) & h_{s-3}(a,\beta,\gamma,\delta) & h_{t-3}(a,\beta,\gamma,\delta) \end{vmatrix} \quad (4)$$

Next let us perform $\text{row}_1+\beta\text{row}_2$, $\text{row}_2+\gamma\text{row}_3$, $\text{row}_3+\delta\text{row}_4$; then $\text{row}_1+\gamma\text{row}_2$, $\text{row}_2+\delta\text{row}_3$; then $\text{row}_1+\delta\text{row}_2$, using (1) each time. The factors previously removed, $\beta-a$, $\gamma-a$, $\delta-a$, ..., form the difference-product $|A(0123)|$, so that writing $h_j(a,\beta,\gamma,\delta)=h_j$ we have now

$$|A(0rst)|/|A(0123)| = \begin{vmatrix} h_0 & h_r & h_s & h_t \\ \cdot & h_{r-1} & h_{s-1} & h_{t-1} \\ \cdot & h_{r-2} & h_{s-2} & h_{t-2} \\ \cdot & h_{r-3} & h_{s-3} & h_{t-3} \end{vmatrix}, \quad . \quad (5)$$

which, if the order of rows and columns be reversed, is seen to be a minor belonging to consecutive rows of $|H|$ in **48**. Let us observe that when the first row is given, or the diagonal suffixes 0, $r-1$, $s-2$, $t-3$, the bialternant (5) is completely determined; also that the sum of the diagonal suffixes $(r-1)+(s-2)+(t-3)$, where $0 \leqslant r-1 \leqslant s-2 \leqslant t-3$, is the degree of the bialternant. The diagonal suffixes thus constitute a *partition* of the degree.

1. The partitions of the integer 4, arranged in ascending order, are $1+1+1+1$, $1+1+2$, $1+3$, $2+2$ and 4. There are therefore five bialternants of the 4th degree, which the reader should write out in full. The last two, for example, are

$$\begin{vmatrix} h_0 & h_3 & h_4 \\ \cdot & h_2 & h_3 \\ \cdot & h_1 & h_2 \end{vmatrix} \text{ and } \begin{vmatrix} h_0 & h_5 \\ \cdot & h_4 \end{vmatrix},$$

and since $h_0 = 1$ the first row and column in these may be deleted.

2. $|A(012t)|/|A(0123)| = h_{t-3}$, $|A(0123t)|/|A(01234)| = h_{t-4}$ and so on.

The determinantal form (5) for a bialternant is due to Jacobi (1841) and can be derived for a general bialternant by the method we have exemplified. We may notice two features in the bialternant : (i) the suffixes of the h_j in its first row are the indices of the alternant in the numerator of the quotient of alternants ; (ii) the sum of the suffixes in every term of the expansion of a bialternant in its determinantal form is equal to the degree, for example, $(r-1)+(s-2)+(t-3)$ in (5). The sum of the suffixes is often called the *weight*, and a polynomial in which each term is of the same weight is said to be *isobaric*.

Dual Form of a Bialternant. By the aid of Jacobi's theorem on minors of the adjugate we may obtain an alternative form for the bialternant as an isobaric determinant like (5), but with elements a_j instead of h_j. For suppose the matrices A and H in **48** (5) to be of order $(t+1) \times (t+1)$. Then the lower left corner element of H is h_t, and the upper right corner element of A' is $(-)^t a_t$. Let rows be named by the highest suffixes of the h_j or the a_j which they contain. The rows of H are therefore named by 0, 1, 2, ..., t and those of A' by the reversed order t, $t-1$, $t-2$, ..., 2, 1, 0. Now the bialternant (5) for example, when transposed, belongs to rows 0, r, s, t of H, columns 0, 1, 2, 3. Hence, by Jacobi's theorem, the complementary bialternant in A' belongs to rows obtained by removing 0, r, s, t from the sequence 0, 1, 2, ..., t and taking the *complements* with respect to t (because of the reversed order of rows in A') of the numbers that remain.

For example, if 0, r, s, t were 0, 1, 3, 6 the numbers not contained are 2, 4, 5 and their complements with respect to 6 are, in ascending order, 1, 2, 4. Hence we have the identity

$$\begin{vmatrix} h_0 & h_1 & h_3 & h_6 \\ \cdot & h_0 & h_2 & h_5 \\ \cdot & \cdot & h_1 & h_4 \\ \cdot & \cdot & h_0 & h_3 \end{vmatrix} = \begin{vmatrix} a_1 & a_2 & a_4 \\ a_0 & a_1 & a_3 \\ \cdot & a_0 & a_2 \end{vmatrix}, \qquad . \qquad . \quad (6)$$

where we have made all elements of the determinant on the

right positive by altering the sign of the first row and second column in the corresponding signed minor of A'.

The proof which we have here sketched in outline from an example is general, and establishes a *dual* or complementary relationship between bialternants expressed in elements h_j or in elements a_j. Sets of integers like 0, 1, 3, 6 and 1, 2, 4 are said to be *bicomplementary* with respect to the highest index, such as 6 here, and the theorem of duality between bialternants may be succinctly expressed thus: the suffixes of elements in the first rows of dual bialternants form bicomplementary sets.

3. Find the sets bicomplementary to (0125), (0235), (025) and (01245) with respect to 5. Write down the corresponding identities between dual bialternants. Interchange h with a and note that all the identities remain true, because of the symmetry of Wronski's relations, **48** (4) and (5).

4. Prove that h_r is equal to a bialternant of order r with all diagonal elements equal to a_1. Here again h and a may be interchanged.

Reciprocal of Alternant Matrix. The reciprocal of $A(012)$ is

$$
\begin{bmatrix} 1 & \alpha & \alpha^2 \\ 1 & \beta & \beta^2 \\ 1 & \gamma & \gamma^2 \end{bmatrix}^{-1} =
$$

$$
\begin{bmatrix} \dfrac{\beta\gamma}{(\alpha-\beta)(\alpha-\gamma)} & \dfrac{\gamma\alpha}{(\beta-\gamma)(\beta-\alpha)} & \dfrac{\alpha\beta}{(\gamma-\alpha)(\gamma-\beta)} \\[2ex] -\dfrac{\beta+\gamma}{(\alpha-\beta)(\alpha-\gamma)} & -\dfrac{\gamma+\alpha}{(\beta-\gamma)(\beta-\alpha)} & -\dfrac{\alpha+\beta}{(\gamma-\alpha)(\gamma-\beta)} \\[2ex] \dfrac{1}{(\alpha-\beta)(\alpha-\gamma)} & \dfrac{1}{(\beta-\gamma)(\beta-\alpha)} & \dfrac{1}{(\gamma-\alpha)(\gamma-\beta)} \end{bmatrix} . \quad (7)
$$

The numerators in the elements of the respective columns are elementary symmetric functions in the arguments α, β, γ with one argument omitted each time. The nature of the polynomials in the denominators of the elements is clear from the example. The reader will be able to extend the result to the case of a general alternant of order $n \times n$.

50. Confluent or Differentiated Alternants

A polynomial $f(x)$ of the nth degree is uniquely determined by the values which it assumes for $n+1$ different values of x. When two or more of these points coincide the polynomial is no longer determinate ; but if we are provided with the values of a sufficient number of derivatives of $f(x)$ at the points of coincidence $f(x)$ again becomes determinate. For example a polynomial of the 4th degree is fully determined by the values of $f(a)$, $f'(a)$, $f''(a)$, $f(\beta)$, $f'(\beta)$, provided that a and β are different. The condition of consistency which underlies the five equations involved is expressed by an alternant matrix in which the elements of certain rows have been differentiated. We shall call such a matrix a *confluent alternant matrix,* and its determinant a *confluent alternant.*

For example, if $f(x) = c_0 + c_1 x + c_2 x^2 + c_3 x^3 + c_4 x^4$. (1) then the expressions for $f(a)$, $f'(a)$, $f''(a)/2!$, $f(\beta)$, $f'(\beta)$ give five equations in the coefficients c_i with matrix

$$\begin{bmatrix} 1 & a & a^2 & a^3 & a^4 \\ . & 1 & 2a & 3a^2 & 4a^3 \\ . & . & 1 & 3a & 6a^2 \\ 1 & \beta & \beta^2 & \beta^3 & \beta^4 \\ . & 1 & 2\beta & 3\beta^2 & 4\beta^3 \end{bmatrix}, \qquad . \quad . \quad . \quad (2)$$

an example of a confluent alternant matrix. The operations on rows are of the type $(d/dx)^k/k!$.

1. Construct the confluent alternant matrices corresponding to $f(a)$, $f'(a)$, $f(\beta)$, $f'(\beta)$ and to $f(a)$, $f'(a)$, $f''(a)/2!$, $f'''(a)/3!$, where $f(x)$ is a cubic polynomial.

Evaluation of Confluent Alternant Determinant. Let us consider

$$|A(01234)| = \begin{vmatrix} 1 & a & a^2 & a^3 & a^4 \\ 1 & \beta & \beta^2 & \beta^3 & \beta^4 \\ 1 & \gamma & \gamma^2 & \gamma^3 & \gamma^4 \\ 1 & \delta & \delta^2 & \delta^3 & \delta^4 \\ 1 & \epsilon & \epsilon^2 & \epsilon^3 & \epsilon^4 \end{vmatrix} \qquad . \quad . \quad (3)$$

in which α, β, γ are going to coalesce to the same value α, and δ and ϵ to the same value δ. Let us apply to rows 1, 2, 3 and to rows 4, 5, separately, the first set of operations used on the alternant of **49** (3). We thus obtain

$$|A(01234)|/\{\Delta(\alpha, \beta, \gamma)\Delta(\delta, \epsilon)\}$$

$$= \begin{vmatrix} 1 & \alpha & \alpha^2 & \alpha^3 & \alpha^4 \\ . & 1 & h_1(\alpha, \beta) & h_2(\alpha, \beta) & h_3(\alpha, \beta) \\ . & . & 1 & h_1(\alpha, \beta, \gamma) & h_2(\alpha, \beta, \gamma) \\ 1 & \delta & \delta^2 & \delta^3 & \delta^4 \\ . & 1 & h_1(\delta, \epsilon) & h_2(\delta, \epsilon) & h_3(\delta, \epsilon) \end{vmatrix}. \qquad (4)$$

Now since by **48** (2) the function $\{(1-\alpha t)(1-\beta t)(1-\gamma t)\}^{-1}$ generates the symmetric function $h_j(\alpha, \beta, \gamma)$ as coefficient of t^j in its expansion, the generating function when $\alpha = \beta = \gamma$ is $(1-\alpha t)^{-3}$, which can be expanded by the binomial theorem. In the same way when $\delta = \epsilon$ the generating function of $h_j(\delta, \epsilon)$ is $(1-\delta t)^{-2}$. Expanding these generating functions we see that when α, β, γ coalesce to α and δ, ϵ coalesce to δ the determinant on the right of (4) becomes

$$\begin{vmatrix} 1 & \alpha & \alpha^2 & \alpha^3 & \alpha^4 \\ . & 1 & 2\alpha & 3\alpha^2 & 4\alpha^3 \\ . & . & 1 & 3\alpha & 6\alpha^2 \\ 1 & \delta & \delta^2 & \delta^3 & \delta^4 \\ . & 1 & 2\delta & 3\delta^2 & 4\delta^3 \end{vmatrix}. \qquad (5)$$

But this is the confluent alternant. Hence its value is

$$\lim_{\substack{\beta, \gamma \to \alpha, \epsilon \to \delta}} \Delta(\alpha, \beta, \gamma, \delta, \epsilon)/\{\Delta(\alpha, \beta, \gamma)\Delta(\delta, \epsilon)\}$$
$$= \lim (\epsilon-\gamma)(\epsilon-\beta)(\epsilon-\alpha)(\delta-\gamma)(\delta-\beta)(\delta-\alpha) = (\delta-\alpha)^6. \qquad (6)$$

This function may be called the *confluent difference-product*. It may be denoted by $\Delta\{(\alpha\alpha\alpha)(\delta\delta)\}$, being regarded as constructed from differences of arguments taken one from each set (α, α, α) and (δ, δ) in reversed order, the zero differences of elements from the same set being excluded.

Following the above lines the reader will show that the value of the general confluent alternant in which n_1 elements coalesce to a_1, n_2 to a_2, ..., and finally n_k to a_k is

$$\Delta\{(a_1, a_1,..., a_1)(a_2, a_2,..., a_2)...(a_k, a_k,..., a_k)\} = \Pi(a_k - a_i)^{n_i n_k}, i < k. \quad (7)$$

Reciprocal of Confluent Alternant Matrix. The problem of finding the reciprocal of a confluent alternant matrix may be approached by interpreting $A^{-1}P^{-1}$ as operations on columns of A^{-1}, where PA is the result of elementary operations (**19**, Ex. 21) on the rows of A. In view of **21**, Ex. 7 and 9, it is seen that if P in premultiplication effects operations $\text{row}_i - \lambda\text{row}_k$, μrow_h, then P^{-1} in postmultiplication effects $\text{col}_k + \lambda\text{col}_i$, $\mu^{-1}\text{col}_h$. Now a confluent alternant is derived (**50** (3), (4), (5)) from a simple alternant A by operations PA like $\text{row}_2 - \text{row}_1$, $(\beta - a)^{-1}\text{row}_2$, $\text{row}_3 - \text{row}_2$, $(\gamma - \beta)^{-1}\text{row}_3$ and so on. Hence the reciprocal confluent alternant $A^{-1}P^{-1}$ may be determined from A^{-1} (for this see **49** (7)) by operations like $\text{col}_1 + \text{col}_2$, $(\beta - a)\text{col}_2$, $\text{col}_2 + \text{col}_3$, $(\gamma - \beta)\text{col}_3$ and so on, and then permitting coalescence of appropriate arguments.

It will be found instructive to prove, both by this method and by direct formation of the reciprocal, that

$$\begin{bmatrix} 1 & a & a^2 \\ \cdot & 1 & 2a \\ 1 & \gamma & \gamma^2 \end{bmatrix}^{-1} = \begin{bmatrix} \dfrac{\gamma(\gamma - 2a)}{(a - \gamma)^2} & \dfrac{\gamma a}{a - \gamma} & \dfrac{a^2}{(\gamma - a)^2} \\ \dfrac{2a}{(a - \gamma)^2} & -\dfrac{\gamma + a}{a - \gamma} & -\dfrac{2a}{(\gamma - a)^2} \\ -\dfrac{1}{(a - \gamma)^2} & \dfrac{1}{a - \gamma} & \dfrac{1}{(\gamma - a)^2} \end{bmatrix}. \quad (8)$$

51. Persymmetric, Circulant and Centrosymmetric Types

A determinant in which all elements in any diagonal at right angles to the principal diagonal are the same was formerly called *orthosymmetric*. Adopting a more modern name we shall call both the determinant and its matrix *persymmetric*. For example the matrix

$$A = \begin{bmatrix} a_1 & a_2 & a_3 & a_4 \\ a_2 & a_3 & a_4 & a_5 \\ a_3 & a_4 & a_5 & a_6 \\ a_4 & a_5 & a_6 & a_7 \end{bmatrix} \qquad . \qquad . \qquad . \quad (1)$$

is persymmetric. Evidently persymmetry of a matrix $A = [a_{ij}]$ is completely characterized by the statement $[a_{ij}] = [a_{i+j-1}]$. Also the number of independent elements in a persymmetric matrix of order $n \times n$ is $2n-1$.

Persymmetric determinants $|A|$ with special elements can often be evaluated more easily by first forming DA, DAD' or LA, LAL', where, taking $n = 4$ for illustration, we have

$$D = \begin{bmatrix} 1 & . & . & . \\ -1 & 1 & . & . \\ 1 & -2 & 1 & . \\ -1 & 3 & -3 & 1 \end{bmatrix}, L = \begin{bmatrix} 1 & . & . & . \\ -\lambda & 1 & . & . \\ \lambda^2 & -2\lambda & 1 & . \\ -\lambda^3 & 3\lambda^2 & -3\lambda & 1 \end{bmatrix}, (2)$$

the coefficients appearing in the rows being the binomial coefficients of successive order. Since $\Delta a_1 = a_2 - a_1$, $\Delta^2 a_1 = a_3 - 2a_2 + a_1$ and so on and $|D|$ and $|L| = 1$, we find the following equivalents for the determinant $|A|$,

$$|DA| = \begin{vmatrix} a_1 & a_2 & a_3 & a_4 \\ \Delta a_1 & \Delta a_2 & \Delta a_3 & \Delta a_4 \\ \Delta^2 a_1 & \Delta^2 a_2 & \Delta^2 a_3 & \Delta^2 a_4 \\ \Delta^3 a_1 & \Delta^3 a_2 & \Delta^3 a_3 & \Delta^3 a_4 \end{vmatrix}, |DAD'| = \begin{vmatrix} a_1 & \Delta a_1 & \Delta^2 a_1 & \Delta^3 a_1 \\ \Delta a_1 & \Delta^2 a_1 & \Delta^3 a_1 & \Delta^4 a_1 \\ \Delta^2 a_1 & \Delta^3 a_1 & \Delta^4 a_1 & \Delta^5 a_1 \\ \Delta^3 a_1 & \Delta^4 a_1 & \Delta^5 a_1 & \Delta^6 a_1 \end{vmatrix}, (3)$$

or the same with Δ_λ^s instead of Δ^s, where Δ_λ is the operation of λ-differencing defined by $\Delta_\lambda a_1 = a_2 - \lambda a_1$. We may express these results in concise form as

$$|a_{i+j-1}| = |\Delta_\lambda^{i-1} a_j| = |\Delta_\lambda^{i+j-2} a_1|. \qquad . \qquad . \quad (4)$$

1. Evaluate

$$\begin{vmatrix} 1 & 3 & 6 & 10 \\ 3 & 6 & 10 & 15 \\ 6 & 10 & 15 & 21 \\ 10 & 15 & 21 & 28 \end{vmatrix}$$

after first constructing a table of successive differences of 1, 3, 6, 10, ..., 21, 28.

2. Evaluate

$$\begin{vmatrix} 1 & n & n_{(2)} & n_{(3)} \\ n & n_{(2)} & n_{(3)} & n_{(4)} \\ n_{(2)} & n_{(3)} & n_{(4)} & n_{(5)} \\ n_{(3)} & n_{(4)} & n_{(5)} & n_{(6)} \end{vmatrix}$$

where $n_{(r)} = n(n-1)(n-2) \ldots (n-r+1)/r!$, by repeated use of the identity $(n+1)_{(r)} - n_{(r)} = n_{(r-1)}$.

Circulant Matrices and Determinants. A matrix of order $n \times n$ of the type

$$C = \begin{vmatrix} a_1 & a_2 & a_3 & a_4 \\ a_4 & a_1 & a_2 & a_3 \\ a_3 & a_4 & a_1 & a_2 \\ a_2 & a_3 & a_4 & a_1 \end{vmatrix}, \qquad \cdot \quad \cdot \quad \cdot \quad (5)$$

illustrated by the case $n = 4$, is called a *circulant matrix*, and its determinant is called a *circulant*. There are only n independent elements. They appear in the first row, and the elements in successive rows are the successive cyclic permutations of these. By inspection C is persymmetric about the secondary diagonal.

If all later rows of C are added to the first row, each element in the resulting first row is equal to $a_1+a_2+\ldots+a_n$. This therefore is a factor of the circulant $|C|$.

This is only one of n factors, the typical factor being $a_1+a_2\omega_j+a_3\omega_j^2+ \ldots +a_n\omega_j^{n-1}$, where ω_j is an nth root of 1. To see that this is so, let us construct $\Omega^{-1}C\Omega$, where Ω is a diagonal matrix of the type exemplified by

$$\Omega = \begin{bmatrix} 1 & & & \\ & \omega & & \\ & & \omega^2 & \\ & & & \omega^3 \end{bmatrix}, \qquad \cdot \quad \cdot \quad \cdot \quad (6)$$

the diagonal elements being the powers of a primitive n^{th} root of 1. The effect, as the reader will verify, is to produce a new circulant in which a_k has been replaced by $a_k\omega^{k-1}$. At the same time $|C|$ is unaltered in value, since $|\Omega|^{-1}|C|\,|\Omega| = |C|$. Hence by adding as before the later rows to the first row it is seen that $a_1+a_2\omega+ \ldots +a_n\omega^{n-1}$ is a factor of $|\Omega^{-1}C\Omega|$ or $|C|$. Thus $|C|$ contains the continued product of n such factors ; and since the leading term derived from the diagonal elements is a_1^n the only remaining factor must be 1. The result in question is thus established.

3. $\begin{vmatrix} a_1 & a_2 \\ a_2 & a_1 \end{vmatrix} = (a_1+a_2)(a_1-a_2).$

4. $\begin{vmatrix} a_1 & a_2 & a_3 \\ a_3 & a_1 & a_2 \\ a_2 & a_3 & a_1 \end{vmatrix} = (a_1+a_2+a_3)(a_1^2+a_2^2+a_3^2-a_1a_2-a_1a_3-a_2a_3)$

$= (a_1+a_2+a_3)(a_1+\omega a_2+\omega^2 a_3)(a_1+\omega^2 a_2+\omega a_3),$

where ω and ω^2 are the complex cube roots of 1.

5. Write down the corresponding factors of the circulant of order 4 and verify by actual construction of their product that it is equal to the circulant.

6. The sum and the product of two circulant matrices, and the reciprocal of a circulant matrix, are circulant.

Centrosymmetric Matrices and Determinants. A matrix which is symmetric about the centre of its array of elements is said to be *centrosymmetric*. Thus if K is centrosymmetric of order $n \times n$ we must have $k_{ij} = k_{n-i+1,\ n-j+1}.$

If we make use of the matrix J (**11**, Ex. 3, 4) which has units in the secondary diagonal and zeros elsewhere, and which in premultiplication reverses the rows of a matrix and in postmultiplication reverses the columns, we may express a centrosymmetric matrix as a partitioned matrix

$$K = \begin{bmatrix} A & B \\ JBJ & JAJ \end{bmatrix} \text{ or } K = \begin{bmatrix} A & a & B \\ b' & a_{mm} & b'J \\ JBJ & Ja & JAJ \end{bmatrix}, \quad (7)$$

according as K is of order $2m \times 2m$ or $(2m-1) \times (2m-1)$. In the second case a is a column vector, b' is a row vector, and a_{mm} is the central element.

In either case a centrosymmetric determinant can be resolved into two determinant factors, for in the first we have

$$\begin{bmatrix} I & J \\ . & I \end{bmatrix} \begin{bmatrix} A & B \\ JBJ & JAJ \end{bmatrix} \begin{bmatrix} I & -J \\ . & I \end{bmatrix} = \begin{bmatrix} A+BJ & . \\ JBJ & JAJ-JB \end{bmatrix}, \quad (8)$$

as the reader should verify, interpreting also the operations as row and column operations upon K. Thus by Laplacian expansion

$$|K| = |A+BJ| \ |JAJ-JB|. \qquad . \qquad (9)$$

In the second case we have

$$\begin{bmatrix} I & . & J \\ . & 1 & . \\ . & . & I \end{bmatrix} \begin{bmatrix} A & a & B \\ b' & a_{mm} & b'J \\ JBJ & Ja & JAJ \end{bmatrix} \begin{bmatrix} I & . & -J \\ . & 1 & . \\ . & . & I \end{bmatrix}$$

$$= \begin{bmatrix} A+BJ & 2a & . \\ b' & a_{mm} & . \\ JBJ & Ja & JAJ-JB \end{bmatrix}, \qquad . \qquad (10)$$

which by Laplacian expansion gives

$$|K| = \begin{vmatrix} A+BJ & 2a \\ b' & a_{mm} \end{vmatrix} \ |JAJ-JB|,. \qquad . \qquad (11)$$

the second factor being of the same form as in (9). The factors here are of orders m and $m-1$.

7. Construct centrosymmetric determinants of orders 2, 3, 4 and 5 with literal elements and factorize them.

52. Dialytic Elimination : Bigradient Matrix

Consider two algebraic equations, such as

$$a_0x^2+a_1x+a_2 = 0, \qquad . \qquad (1)$$
$$b_0x^3+b_1x^2+b_2x+b_3 = 0, \qquad . \qquad (2)$$

in the general case of degrees m and n. Let it be asked

what condition the coefficients a_j and b_j must satisfy so that the equations may have a root in common. Multiplying (1) by 1, x, x^2 and (2) by 1, x in turn, we obtain five homogeneous equations in 1, x, x^2, x^3, x^4. The condition for their consistency is that their determinant

$$\begin{vmatrix} a_0 & a_1 & a_2 & . & . \\ . & a_0 & a_1 & a_2 & . \\ . & . & a_0 & a_1 & a_2 \\ . & b_0 & b_1 & b_2 & b_3 \\ b_0 & b_1 & b_2 & b_3 & . \end{vmatrix} = 0. \quad . \quad . \quad . \quad (3)$$

Such a determinant, or such a matrix, is said to be *bigradient*. (We have conformed to custom in making the persymmetry in the a_j and b_j run in two different slopes.) In the general case the bigradient is of order $m+n$, with n rows of elements a_j sloping down persymmetrically, and m rows of elements b_j sloping up. The process of eliminating x in this way from two polynomials is due to Sylvester, and is called *dialytic elimination*.

1. Express by means of a bigradient the condition that the quadratic equation $a_0x^2+a_1x+a_2 = 0$ should have a double root. Evaluate the bigradient.

2. In the same way find the condition that the general cubic equation should have a repeated root. (The second polynomial in each of these examples will be the derivative of the first.)

53. Continuant Matrices and Continuants

A matrix C_n of order $n \times n$ which has zero elements everywhere except in the principal diagonal, the superdiagonal directly above it, and the subdiagonal directly below, is called a *continuant matrix*, and its determinant a *continuant*. For example

$$\begin{bmatrix} c_0 & d_1 & . & . \\ b_1 & c_1 & d_2 & . \\ . & b_2 & c_2 & d_3 \\ . & . & b_3 & c_3 \end{bmatrix} \text{ and } \begin{bmatrix} c_0 & d_1 & . & . \\ -1 & c_1 & d_2 & . \\ . & -1 & c_2 & d_3 \\ . & . & -1 & c_3 \end{bmatrix} \quad . \quad (1)$$

are continuant matrices. In the second of these we have divided the subdiagonal elements by suitable factors to make them all -1. This is the usual convention.

The name *continuant* is due to the intimate relation which the continuant matrix, in the above standard form, bears to the theory of continued fractions. Consider for example a sequence of continued fractions, in the usual notation,

$$\frac{1}{c_0+} \frac{d_1}{c_1+} \frac{d_2}{c_2+} \cdots \frac{d_j}{c_j} = \frac{p_j}{q_j}, \quad j = 0, 1, 2, \ldots, n, . \quad (2)$$

the first n of these being convergents to the final one. It is not difficult to show that the numerators p_j satisfy the set of recurrence relations

$$p_0 = 1, \; p_1 = c_1 p_0, \; \ldots, \; p_j = c_j p_{j-1} + d_j p_{j-2}, \; j > 1, \quad (3)$$

while the denominators satisfy the very similar relations

$$q_0 = c_0, \; q_1 = c_1 q_0 + d_1, \; \ldots, \; q_j = c_j q_{j-1} + d_j q_{j-2}, \; j > 1. \quad (4)$$

Now the continuants $|C_j|$ satisfy the same relations. For expanding $|C_j|$ according to its last row and column, and noting the initial values $|C_0|$ and $|C_1|$, we have

$$|C_0| = c_0, \; |C_1| = c_1|C_0| + d_1, \; |C_j| = c_j|C_{j-1}| + d_j|C_{j-2}|, \quad (5)$$

so that the continuants $|C_j|$ are equal to q_j, the denominators of convergents of the associated continued fraction.

The numerators of the convergents are also continuants. For the cofactor of the leading element c_0 in $|C_j|$, let us say $C|_{j;\,0}|$, is a continuant. Expanding it by a similar Cauchy expansion and noting initial values, we have

$$|C_{1;\,0}| = c_1, \; |C_{2;\,0}| = c_2|C_{1;\,0}| + d_2, \; |C_{j;\,0}| = c_j|C_{j-1;\,0}| + d_j|C_{j-2;\,0}|, \quad (6)$$

so that $|C_{j;\,0}| = p_j$. Thus the numerator of any convergent is the cofactor of the leading element in the continuant representing the denominator.

Hence the continued fraction is equal to $|C_{n,0}|/|C_n|$, which is the leading element in the reciprocal matrix C_n^{-1}. This fundamental result connects the theory of continuant matrices with the theory of continued fractions.

   1. A continuant in standard form and with elements $c_i = 1$, considered as a polynomial, is prime to its leading minor of order n, namely the cofactor of its last element c_n.

   2. A continuant in standard form, with unit elements in the superdiagonal and integer elements in the diagonal, is prime to the cofactor of its first element.

54. Jacobian, Hessian and Wronskian Matrices

There is an important class of matrices the elements of which are functions and their derivatives.

Consider n functions u_j of n independent variables x_i, the u_j possessing first partial derivatives within a certain region. Let us suppose first that the u_j obey a relation

$$\phi(u_1, u_2 ..., u_n) \equiv \phi = 0, \qquad \qquad (1)$$

where ϕ possesses first derivatives $\partial\phi/\partial u_j$.

Differentiating ϕ with respect to the x_i we obtain

$$\frac{\partial\phi}{\partial x_i} = \frac{\partial\phi}{\partial u_1}\frac{\partial u_1}{\partial x_i} + \frac{\partial\phi}{\partial u_2}\frac{\partial u_2}{\partial x_i} + ... + \frac{\partial\phi}{\partial u_n}\frac{\partial u_n}{\partial x_i} = 0. \qquad (2)$$

These are n homogeneous equations in the $\partial\phi/\partial u_j$, and for their consistency the determinant $|\partial u_j/\partial x_i|$ must vanish. This determinant is called the *Jacobian* of the functions u_j with respect to the variables x_i, and is often denoted by

$$\frac{\partial(u_1, u_2, ..., u_n)}{\partial(x_1, x_2, ..., x_n)}. \qquad \qquad (3)$$

It will be convenient to transpose the matrix of the system (2) and to regard $[\partial u_i/\partial x_j]$, or briefly $[\partial u/\partial x]$, as the *Jacobian matrix* of the vector u with respect to the vector x. We may also consider m functions of n variables, with *Jacobian* matrix of order $m \times n$.

   1. If $y = Ax$, then $[\partial y/\partial x] = A$. Thus $[\partial x/\partial x] = I$.

   2. If $v = Au$, then $[\partial v/\partial x] = A[\partial u/\partial x]$.

   3. If $x = Ay$ the column vectors $\{\partial u/\partial x_i\}$ and $\{\partial u/\partial y_i\}$ are related according to $\{\partial u/\partial y_i\} = A'\{\partial u/\partial x_i\}$. In fact $[\partial u/\partial y] = [\partial u/\partial x]\,A$.

4. The reader should verify the above results for cases of low order written in full, and for special functions. Jacobians of chosen sets of functions should also be evaluated ; e.g. $x_1^2 + x_2^2 + x_3^2$, $x_1 x_2 + x_2 x_3 + x_3 x_1$, $x_1 + x_2 + x_3$.

Change of Variables. Consider now m functions u_i of n independent variables y_j, these y_j being functions of n independent x_k. We have then $[\partial u/\partial y]$ and $[\partial y/\partial x]$, of orders $m \times n$ and $n \times n$. It is fundamental that

$$[\partial u/\partial y]\ [\partial y/\partial x] = [\partial u/\partial x]. \qquad . \qquad . \qquad (4)$$

In fact the $(i, j)^{th}$ element of the product on the left is

$$\frac{\partial u_i}{\partial y_1}\ \frac{\partial y_1}{\partial x_j} + \frac{\partial u_i}{\partial y_2}\ \frac{\partial y_2}{\partial x_j} + \ldots + \frac{\partial u_i}{\partial y_n}\ \frac{\partial y_n}{\partial x_j}, \qquad . \qquad (5)$$

which, by the differential calculus, is equal to $\partial u_i/\partial x_j$.

It follows at once that $[\partial y/\partial x]$ and $[\partial x/\partial y]$ are reciprocal matrices, since by (4) their product is I.

The reason for the notation (3) for a Jacobian now appears. For taking determinants of square Jacobian matrices in (5) we have $|\partial u/\partial y|\ |\partial y/\partial x| = |\partial u/\partial x|$, which in the case $n = 1$ is the familiar formula for change of variable in a derivative. In fact a Jacobian might be described as a multiple first differential coefficient of n functions with respect to n variables.

This description is warranted by the fact (see for example Gillespie, *Integration*, § 16) that when a multiple differential $dx_1 dx_2 \ldots dx_n$ is transformed to new independent variables y_j the new differential element is $|\partial x/\partial y|\ dy_1 dy_2 \ldots dy_n$. The proof of this fundamental result is outside our scope, but the reader may consider the case of two variables, the transformation from rectangular coordinates x, y to other coordinates u, v, for example to polar coordinates r, θ. Through a point P in the (x, y) plane let curves $u = $ const., $v = $ const. be drawn, and on them let neighbouring points Q and R be taken respectively. Let the corresponding curves through Q and R intersect in

I

S. Then apart from higher infinitesimals, the coordinates of Q, R, S if P be taken as origin (0, 0), are

$$\left(\frac{\partial x}{\partial v}dv, \frac{\partial y}{\partial v}dv\right), \left(\frac{\partial x}{\partial u}du, \frac{\partial y}{\partial u}du\right), \left(\frac{\partial x}{\partial u}du+\frac{\partial x}{\partial v}dv, \frac{\partial y}{\partial u}du+\frac{\partial y}{\partial v}dv\right).$$

Then the area of the elementary triangle PQR (**19**, Ex. 17) is found to be

$$\frac{1}{2}\begin{vmatrix} \dfrac{\partial x}{\partial u} & \dfrac{\partial y}{\partial u} \\ \dfrac{\partial x}{\partial v} & \dfrac{\partial y}{\partial v} \end{vmatrix} dudv = \frac{1}{2}\frac{\partial(x,\ y)}{\partial(u,\ v)}dudv,\ . \qquad (7)$$

and by taking S as origin we find the same value for SQR. Thus taking both triangles we have the element of area PQSR.

Similar reasoning applies in three dimensions. The element of volume enclosed by neighbouring contours u = const., v = const., w = const. may be dissected into six equal elementary tetrahedra, the total volume being thus $dudvdw$ multiplied by the Jacobian.

Test of Functional Dependence. We have seen that functional dependence implies the vanishing of a Jacobian. The converse theorem is also true and is an important test of dependence, but the accurate proof of it exceeds our limits of space and in any case belongs more to function theory.

5. Test by the Jacobian the functional dependence of

$$x_1+x_2+x_3,\ x_1^2+x_2^2+x_3^2-x_1x_2-x_1x_3-x_2x_3,$$
$$x_1^3+x_2^3+x_3^3-3x_1x_2x_3.$$

Hessian Matrix and Hessian. The Jacobian matrix, with respect to the x_i, of the first derivatives $\partial u/\partial x_j$ of a function u is called the *Hessian matrix* of u, its determinant the *Hessian* of u. The Hessian matrix is therefore

$$H = \left[\frac{\partial^2 u}{\partial x_i \partial x_j}\right]. \qquad . \qquad . \qquad (8)$$

In a wide class of cases it is immaterial in which order of differentiation the second partial derivatives are formed, and for such functions u the Hessian matrix is symmetric, $H' = H$.

If the variables x are transformed to new variables y by $x = Ay$, the $\partial u/\partial x_i$ are transformed (Ex. 3) according to $\{\partial u/\partial y_i\} = A'\{\partial u/\partial x_i\}$. Hence, applying the second part of Ex. 3, we obtain the Jacobian matrix of the $\partial u/\partial y_j$ with respect to the y_j, in fact the Hessian matrix of u with respect to the y_j, as $A'HA$. The Hessian matrix thus undergoes congruent transformation, and the Hessian $|H|$ is multiplied by $|A|^2$. This is usually expressed by saying that the Hessian is a *covariant* of u.

6. Let the x_j be transformed to y_j by a general transformation of Jacobian matrix $M = [\partial x/\partial y]$. Show that the new H is $M'HM + P$, where P is a certain matrix product.

7. The Hessian matrix of a quadratic form $x'Ax$ is $2A$.

8. Evaluate the Hessian of $ax_1^3 + 3bx_1^2x_2 + 3cx_1x_2^2 + dx_2^3$, and find also the Hessian of $(ax_1 + bx_2)^3$.

9. In the Taylor expansion of $u(x_1, x_2, ..., x_n)$ about the origin $\{0, 0 ..., 0\}$ the term involving second partial derivatives of u is the quadratic form $x'Hx/2!$.

Wronskian Matrix and Wronskian. Before considering the Wronskian we may note how to differentiate a determinant $|A|$ with respect to x, when each element is a function of x. In virtue of the rule

$$\frac{d}{dx}(f_1f_2f_3) = \frac{df_1}{dx}f_2f_3 + \frac{df_2}{dx}f_1f_3 + \frac{df_3}{dx}f_1f_2,$$

each term in the expansion of $|A|$ gives rise, on being differentiated, to n terms, obtained by differentiating each factor a_{ij} in turn. Thus $d|A|/dx$ contains $n \cdot n!$ terms. The aggregate of $n!$ of these terms having differentiated elements belonging to the ith row forms a determinant obtained from $|A|$ by writing da_{ij}/dx for a_{ij} in that row. Thus $d|A|/dx$ is the sum of n determinants, obtained from $|A|$ by differentiating each row in turn.

10. Write out and verify this result for determinants of the second and third orders.

The *Wronskian matrix* of n functions u_i of an independent variable x is defined by

$$W = \left[\frac{d^{i-1} u_j}{dx^{i-1}} \right]. \qquad . \qquad . \qquad . \quad (9)$$

For example, the Wronskian of u_1, u_2, u_3 is

$$W = \begin{bmatrix} u_1 & u_2 & u_3 \\ \dfrac{du_1}{dx} & \dfrac{du_2}{dx} & \dfrac{du_3}{dx} \\ \dfrac{d^2u_1}{dx^2} & \dfrac{d^2u_2}{dx^2} & \dfrac{d^2u_3}{dx^2} \end{bmatrix}. \qquad . \qquad . \quad (10)$$

The chief property of the Wronskian determinant or *Wronskian* is that it vanishes when the functions satisfy a relation of linear dependence with nonzero constant coefficients. For if we have

$$a_1u_1 + a_2u_2 + a_3u_3 = 0$$

we have also two further equations of the same form in the first and second derivatives of u_1, u_2, u_3. The condition of consistency of these equations in a_1, a_2, a_3 is $|W| = 0$.

The converse theorem, that if the Wronskian is zero the functions obey a non-trivial linear relation, is true under certain restrictions.

11. The derivative of a Wronskian is obtained from the Wronskian by differentiating its last row.

12. Prove that the Wronskian of 1, x, x^2, ..., x^{n-1} is a confluent alternant multiplied by a certain constant factor.

13. Evaluate the Wronskians of 1, x^2, x^4 and of x^2, x^3, x^4. Construct other sets of simple functions and evaluate their Wronskians.

Additional Examples

1. Prove that
$$\begin{vmatrix} 1 & 1 & 1 \\ b+c & c+a & a+b \\ bc & ca & ab \end{vmatrix} = -\varDelta(a, b, c).$$

2. Prove that
$$\begin{vmatrix} x & a & a & a \\ a & x & a & a \\ a & a & x & a \\ a & a & a & x \end{vmatrix} = (x+3a)(x-a)^3$$

and generalize this result.

[Put $x = a+y$ and expand diagonally (**37**) in powers of y.]

3. Evaluate similarly, by suitable substitutions,

$$\begin{vmatrix} x & a & a & a \\ a & y & a & a \\ a & a & z & a \\ a & a & a & t \end{vmatrix}, \quad \begin{vmatrix} x & a & a & a \\ b & x & b & b \\ c & c & x & c \\ d & d & d & x \end{vmatrix}, \quad \begin{vmatrix} x & a & a & a \\ b & y & b & b \\ c & c & z & c \\ d & d & d & t \end{vmatrix}.$$

4. If x is a column vector of n elements, show that

$$|\lambda I - xx'| = \lambda^n - (x'x)\lambda^{n-1}.$$

[A determinant of order n and of the form $|D+M|$, where D is a diagonal matrix and M is of very low rank, is most readily evaluated by diagonal expansion (**37**) in terms of the elements of D and cofactors in M.]

5. Prove that
$$\begin{vmatrix} 1 & t & t^2 & t^3 \\ 1 & 1 & 1 & 1 \\ 1 & 2 & 3 & 4 \\ 1 & 2^2 & 3^2 & 4^2 \end{vmatrix} \div \begin{vmatrix} 1 & 1 & 1 \\ 1 & 2 & 3 \\ 1 & 2^2 & 3^2 \end{vmatrix}$$

has the value $(1-t)^3$, and generalize this result.

6. Prove that
$$\begin{vmatrix} 1 & t & t^2 \\ \cos n\theta & \cos(n+1)\theta & \cos(n+2)\theta \\ \sin n\theta & \sin(n+1)\theta & \sin(n+2)\theta \end{vmatrix} \div \sin\theta$$

has the value $1 - 2t\cos\theta + t^2$.

7. Prove that

$$\begin{vmatrix} 1 & t & t^2 & t^3 & t^4 \\ \cos n\theta & \cos (n+1)\theta & \cos (n+2)\theta & \cos (n+3)\theta & \cos (n+4)\theta \\ \sin n\theta & \sin (n+1)\theta & \sin (n+2)\theta & \sin (n+3)\theta & \sin (n+4)\theta \\ . & \cos (n+1)\theta & 2\cos (n+2)\theta & 3\cos (n+3)\theta & 4\cos (n+4)\theta \\ . & \sin (n+1)\theta & 2\sin (n+2)\theta & 3\sin (n+3)\theta & 4\sin (n+4)\theta \end{vmatrix}$$

contains the factor $(1 - 2t \cos \theta + t^2)^2$, and obtain the remaining factor. Generalise the result.

8. The determinant $|A| = | (a_i - b_j)^{-1}|$ of order n is called Cauchy's double alternant. Show that it is equal to

$$(-)^{\frac{1}{2}n(n-1)}\Delta(a_1, a_2, ..., a_n)\Delta(b_1, b_2, ..., b_n)\Big/\prod_{i, j}(a_i - b_j),$$

where $\Delta(a_1, a_2, ..., a_n)$ denotes (**19**, Ex. 11) a difference-product. [To fix ideas, write down the case of the 3rd order. Clear of fractions by multiplying each row by the continued product of the denominators $a_i - b_j$ in that row. The first row of the resulting determinant will then have the appearance

$$[(a_1 - b_2)(a_1 - b_3) \quad (a_1 - b_1)(a_1 - b_3) \quad (a_1 - b_1)(a_1 - b_2)].$$

This determinant is evidently a polynomial of degree $n(n-1)$ in the a_i, b_j. It vanishes when $a_i = a_j$, also when $b_i = b_j$, $i \neq j$. Hence it contains the two difference-products as factors. Their combined degree is $n(n-1)$. Hence the remaining factor is numerical. To find it put each $a_i = b_i$. Notice that all non-diagonal elements then vanish, and that the continued product of all diagonal elements give the difference-product twice ; but to each $a_i - a_j$ there is also an $a_j - a_i$.]

9. Prove that if $P = [(i+j-1)^{-1}]$, of order $n \times n$, then $|P| = \{1!\ 2!\ 3!\ ...\ (n-1)!\}^4/\{1!\ 2!\ 3!\ ...\ (2n-1)!\}$.
[In Cauchy's double alternant put $a_i = n+1, n+2, ..., 2n$, $b_j = n, n-1, ...\ 2, 1$. Test the result on $n = 3, n = 4$.]

10. Prove that if $P = [\{(i+j-1)!\}^{-1}]$, of order $n \times n$, then $|P| = (-)^{\frac{1}{2}n(n-1)}\{1!\ 2!\ ...\ (n-1)!\}^2/\{1!\ 2!\ ...\ (2n-1)!\}$.
[Multiply the columns by $n!$, $(n+1)!$, ..., $(2n-1)!$ respectively. Note that an alternant then appears, with rows reversed.]

11. Prove that
$$\begin{vmatrix} p & a & a & a \\ b & q & a & a \\ b & b & r & a \\ b & b & b & s \end{vmatrix} = \{bf(a) - af(b)\}/(b-a),$$
where $f(x) = (p-x)(q-x)(r-x)(s-x)$.

12. Prove that the Hessian of the circulant
$$\begin{vmatrix} a & b & c \\ c & a & b \\ b & c & a \end{vmatrix}$$
with respect to a, b, c is equal to -2×3^3 times

the circulant. Prove also that the Hessian of the circulant of the 4th order is equal to 3×4^4 times the square of the circulant.

13. Prove that
$$\frac{1}{n!} \begin{vmatrix}{}^+ & & & & {}^+ \\ 0 & 1 & 1 & \dots & 1 \\ 1 & 0 & 1 & \dots & 1 \\ 1 & 1 & 0 & \dots & 1 \\ \cdot & \cdot & \cdot & \cdot & \cdot \\ 1 & 1 & 1 & \dots & 0 \end{vmatrix}$$

$$= 1 - \frac{1}{1!} + \frac{1}{2!} - \frac{1}{3!} + \frac{1}{4!} - \dots + (-)^n \frac{1}{n!},$$

where the permanent is of the nth order.

[Instead of the zero in the diagonal write $1 + \lambda$. Perform diagonal expansion (**37**) in powers of λ, noting that a permanent of order k with unit elements is equal to $k!$.

The permanent enumerates the terms in the expansion of a permanent (or of a determinant) of order n, which do not contain any diagonal element a_{ii}. Thus the example gives the solution of the "problème des rencontres," *e.g.* the probability of putting n letters in n addressed envelopes, so that no letter (the ith, say) is in its correct i^{th} envelope.]

14. If
$$A = \begin{bmatrix} a_1 & a_2 & a_3 & a_4 \\ 1 & \cdot & \cdot & \cdot \\ \cdot & 1 & \cdot & \cdot \\ \cdot & \cdot & 1 & \cdot \end{bmatrix},$$
then $|A - \lambda I| = \lambda^4 - a_1 \lambda^3 - a_2 \lambda^2 - a_3 \lambda - a_4$. Extend this to the case where A is of order $n \times n$.

15. If the roots λ_i of $|A - \lambda I| = 0$ in Ex. 12 are all distinct, and if

$$L = \begin{bmatrix} \lambda_1^3 & \lambda_2^3 & \lambda_3^3 & \lambda_4^3 \\ \lambda_1^2 & \lambda_2^2 & \lambda_3^2 & \lambda_4^2 \\ \lambda_1 & \lambda_2 & \lambda_3 & \lambda_4 \\ 1 & 1 & 1 & 1 \end{bmatrix},$$

an alternant matrix, prove that $AL = L\Lambda$, where Λ is the diagonal canonical form (37, Ex. 7) of A.   Hence $L^{-1}AL = \Lambda$. Generalize this result.

16. If, in Ex. 15, $\lambda_1 = \lambda_2 = \lambda_3 \neq \lambda_4$, and (50) if

$$L = \begin{bmatrix} \lambda_1^3 & 3\lambda_1^2 & 3\lambda_1 & \lambda_4^3 \\ \lambda_1^2 & 2\lambda_1 & 1 & \lambda_4^2 \\ \lambda_1 & 1 & \cdot & \lambda_4 \\ 1 & \cdot & \cdot & 1 \end{bmatrix}, \quad \Lambda = \begin{bmatrix} \lambda_1 & 1 & \cdot & \cdot \\ \cdot & \lambda_1 & 1 & \cdot \\ \cdot & \cdot & \lambda_1 & \cdot \\ \cdot & \cdot & \cdot & \lambda_4 \end{bmatrix},$$

prove that $AL = L\Lambda$, and so $L^{-1}AL = \Lambda$.

[Here Λ is an example of the non-diagonal canonical form alluded to in 37, Ex. 7, for a special matrix A with multiple latent roots.]

17. If $A = \begin{bmatrix} a_1 & a_2 & a_3 & a_4 \\ 1 & \cdot & \cdot & \cdot \\ \cdot & 1 & \cdot & \cdot \\ \cdot & \cdot & 1 & \cdot \end{bmatrix}$, $H = \begin{bmatrix} \cdot & \cdot & \cdot & -1 \\ \cdot & \cdot & -1 & a_1 \\ \cdot & -1 & a_1 & a_2 \\ -1 & a_1 & a_2 & a_3 \end{bmatrix}$,

prove that $HAH^{-1} = A'$.   Observe also that HA is symmetric. Generalize these results.

18. If x_i, y_i, $i = 1, 2, ..., n$, are complex numbers, not all zero in each case, establish the " Schwarzian " inequality

$$\Sigma |x_i y_i|^2 < (\Sigma |x_i|^2) (\Sigma |y_i|^2).$$

[Regard x_i, y_i as elements of vectors x and y, and construct $\lambda x + y$, where λ is a real scalar.   Then the Hermitian form $(\lambda \bar{x} + \bar{y})' (\lambda x + y)$, being a sum of squared moduli, is necessarily positive.   Viewing it as a quadratic in λ, examine its discriminant.]

19. A Hermitian form $\bar{x}'Ax$ which assumes, for $x \neq 0$, exclusively positive values is said to be *positive definite*, and so is its matrix A. For example if H is non-singular then $\bar{x}'\bar{H}'Hx$, being a sum of squared moduli of elements of Hx, is positive definite ; thus $\bar{H}'H$ is positive definite.

Prove that if $\phi(z)$ is a function real and positive in the real range $z = a$ to $z = b$, and μ_j is the " moment "

$$\mu_j = \int_a^b z^j \phi(x) \, dz,$$

then the persymmetric moment matrices $M = [\mu_{i+j-1}]$ of every order are positive definite.

[The quadratic form $\int_a^b (\lambda_0 + \lambda_1 z + \lambda_2 z^2 + \ldots + \lambda_k z^k)^2 \phi(z) dz$ in the real $\lambda_0, \lambda_1, \ldots, \lambda_k$ is necessarily positive. Its matrix expression is $\lambda' M \lambda$.]

20. Prove that if A and B are Hermitian matrices of order $n \times n$, and B is positive definite, then the roots of $|A - \lambda B| = 0$ are real.

[*Cf.* **30**, Ex. 10. There exists $q \neq 0$ such that $Aq = \lambda Bq$. Premultiply by $\bar{q}'$.]

21. A matrix Q of order $n \times n$, $n > 2$, is of the form λM, where λ is scalar and M has unit elements everywhere except in the diagonal, which has elements μ. Determine λ and μ so that Q may be orthogonal. Write down the matrices Q of orders 3×3, 4×4.

22. The vector $x = \{x_1 \ x_2\}$ is transformed to $Hx = y = \{y_1 \ y_2\}$. Find what matrix transforms $\{x_1^2 \ \sqrt{2}x_1x_2 \ x_2^2\}$ to $\{y_1^2 \ \sqrt{2}y_1y_2 \ y_2^2\}$. Denoting this matrix by $H^{[2]}$, prove that $(KH)^{[2]} = K^{[2]}H^{[2]}$. Prove also that $|H^{[2]}| = |H|^3$.

[Denote $[x_1^2 \ \sqrt{2}x_1x_2 \ x_2^2]$ by $x^{[2]}$. Then $H^{[2]}$ is defined by $y^{[2]} = (Hx)^{[2]} = H^{[2]}x^{[2]}$, where x is arbitrary. Consider $z = Ky$. Then

$$z^{[2]} = (KHx)^{[2]} = (KH)^{[2]}x^{[2]}$$

while also $(KHx)^{[2]} = K^{[2]}(Hx)^{[2]} = K^{[2]}H^{[2]}x^{[2]}$.

Hence the result, x being arbitrary. There is a corresponding result for vectors of any order, and derived power and product vectors of any degree.]

23. If H in Ex. 22 is unitary, so is $H^{[2]}$.

[The generalization of this has a geometrical interpretation. If, for example, the point $\{x_1\, x_2\, x_3\}$ in 3-dimensional Cartesian space undergoes orthogonal transformation H, the point

$$\{x_1^3 \ \sqrt{3}x_1^2x_2 \ \sqrt{3}x_1^2x_3 \ \sqrt{3}x_1x_2^2 \ \sqrt{6}x_1x_2x_3 \ \ldots \ \sqrt{3}x_2x_3^2\}$$

in 10-dimensional space undergoes orthogonal transformation $H^{[3]}$.]

24. With the notation of Ex. 22, prove the relation adj $H^{[2]} = |H|$ (adj $H)^{[2]}$.

25. Given that $\displaystyle\int_{-\infty}^{\infty} e^{-\frac{1}{2}z^2}\, dz = \sqrt{2\pi}$, show that

$$\int\!\!\!\int\!\!\!\int_{-\infty}^{\infty} \ldots \exp\,(-\tfrac{1}{2}x'Ax)dx_1 dx_2 \ldots dx_n = (2\pi)^{\frac{1}{2}n}\ |A|^{-\frac{1}{2}},$$

where $x'Ax$ is a positive definite quadratic form in the x_i.

[First reduce A by orthogonal transformation (**37**, Ex. 9) to $H'AH = \Lambda$. Since A is positive definite, all the λ_i are positive. Hence $\Lambda^{\frac{1}{2}}$ is a real diagonal matrix. Putting $H\Lambda^{-\frac{1}{2}} = K$, we thus have $K'AK = I$. Also $|K|^2|A| = 1$, $|K| = |A|^{-\frac{1}{2}}$. Transform by $x = Ky$. The Jacobian is $|K|$, and so (54) $dx_1 dx_2 \ldots dx_n = |A|^{-\frac{1}{2}}\, dy_1\, dy_2 \ldots dy_n$. At the same time $x'Ax$ becomes $y'y$, a sum of squares. Repeated integration now gives the result.]

26. To "complete" a quadratic form $x'Ax + 2y'x$, as one "completes the square" in elementary algebra: we have $x'Ax + 2y'x = (x + A^{-1}y)'A(x + A^{-1}y) - y'A^{-1}y$. Apply this to show that

$$(2\pi)^{-\frac{1}{2}n}|A|^{\frac{1}{2}}\int\!\!\!\int\!\!\!\int_{-\infty}^{\infty} \ldots \exp\,(-\tfrac{1}{2}x'Ax + \beta'x)\, dx_1\, dx_2 \ldots dx_n$$

$$= \exp\,(\tfrac{1}{2}\beta'A^{-1}\beta),$$

where A is positive definite.

[Complete the quadratic form and change to new variables $x - A^{-1}\beta$. The limits and differential element remain unchanged.]

27. If $A \equiv \begin{bmatrix} A_1 & B_1 \\ A_2 & B_2 \end{bmatrix}$ is a partitioned matrix such that

A_1 and B_2 are nonsingular, show that

$$\begin{bmatrix} A_1 & B_1 \\ A_2 & B_2 \end{bmatrix}^{-1} = \begin{bmatrix} (A_1 - B_1 B_2^{-1} A_2)^{-1} & (B_1 B_2^{-1} A_2 - A_1)^{-1} B_1 B_2^{-1} \\ (A_2 A_1^{-1} B_1 - B_2)^{-1} A_2 A_1^{-1} & (B_2 - A_2 A_1^{-1} B_1)^{-1} \end{bmatrix},$$

and that if A_2 and B_1 are also nonsingular, this becomes

$$\begin{bmatrix} (A_1 - B_1 B_2^{-1} A_2)^{-1} & (A_2 - B_2 B_1^{-1} A_1)^{-1} \\ (B_1 - A_1 A_2^{-1} B_2)^{-1} & (B_2 - A_2 A_1^{-1} B_1)^{-1} \end{bmatrix}.$$

[Partition A^{-1} and multiply out $AA^{-1} = I$, all three matrices being partitioned. This gives two pairs of linear matrix equations in the four submatrices of A^{-1}, to be solved by elimination, with due precautions.]

28. Denoting row and column number by s and t, show that the matrix H of order $n \times n$ and defined by

$$[h_{st}] = \frac{1}{\sqrt{n}} \left[e^{is(t-1)\theta} \right],$$

where $i = \sqrt{-1}$, $n\theta = 2\pi$, is unitary.

[It follows that if $v = Hu$, then $u = \bar{H}'v$. In ordinary notation, if

$$v_s = \frac{1}{\sqrt{n}} \sum_{t=0}^{n-1} u_t e^{ist\theta} \text{ then } u_t = \frac{1}{\sqrt{n}} \sum_{s=0}^{n-1} v_s e^{-ist\theta},$$

an example of " Fourier " reciprocity.]

29. If $K = \begin{bmatrix} \cdot & 1 \\ -1 & \cdot \end{bmatrix}$, show that $e^{\theta K} = \begin{bmatrix} \cos\theta & \sin\theta \\ -\sin\theta & \cos\theta \end{bmatrix}$,

where θ is a scalar. Show also that if H is a skew Hermitian matrix, e^H is a unitary matrix.

30. The matrices

$$I = \begin{bmatrix} 1 & \cdot \\ \cdot & 1 \end{bmatrix}, K = \begin{bmatrix} \cdot & 1 \\ -1 & \cdot \end{bmatrix}, L = \begin{bmatrix} i & \cdot \\ \cdot & -i \end{bmatrix}, M = \begin{bmatrix} \cdot & -i \\ -i & \cdot \end{bmatrix},$$

where $i = \sqrt{-1}$, possess the properties $K^2 = L^2 = M^2 = -I$, $KL = M = -LK$, and so on. These are the relations satisfied by the four units, usually written 1, i, j, k, of the algebra of quaternions. There is thus a representation of this algebra by complex matrices of order 2×2. But the units 1, i of the algebra of complex numbers themselves possess (9, Ex. 11) the 2×2 real matrix representation

$$\begin{bmatrix} 1 & . \\ . & 1 \end{bmatrix} \text{ and } \begin{bmatrix} . & 1 \\ -1 & . \end{bmatrix}.$$

Hence, replacing each 1 and i in I, K, L, M by these respective submatrices, we have a 4×4 real representation of quaternion algebra.

INDEX

141

Addition $\quad A + B \equiv [a_{ij}] + [b_{ij}] = [a_{ij} + b_{ij}]$

Multiplication by a Scalar $\quad \lambda A = \lambda[a_{ij}] = [\lambda a_{ij}]$

Multiplicat^n by a Scalar matrix $\quad AB = kB \quad$ where

$\quad k =$ diag. elements of scalar matrix A.

$\therefore \underline{I} B = BI = B \quad$ for here $k = 1$.

Transposed matrix A' of $A \quad$ If $A \equiv [a_{ij}]$

$\quad\quad\quad\quad\quad$ Then $A' \equiv [a_{ji}]$

Reversal rule for products $\quad (AB)' = B'A' \quad$ (not $A'B'$)

$\quad\quad\quad\quad\quad\quad\quad\quad (BA)' = A'B' \quad$ (not $B'A'$)

Complex conjugate of matrix $\bar{A} \quad$ If $A = [a_{ij}]$

$\quad\quad\quad$ then $\bar{A} = [\overline{a_{ij}}] \quad\quad$ (eg $\overline{x+iy} = x-iy$)